INTEGRATED TELECOMMUNICATIONS MANAGEMENT SOLUTIONS

IEEE Press
445 Hoes Lane, P.O. Box 1331
Piscataway, NJ 08855-1331

Technical Reviewers

Larry Bernstein, *AT&T (retired)*
Ray Pyle, *Blue Cross/Blue Shield, Farmington Hills, MI*
Lakshmi Raman, *Telcordia Technologies, Inc., Red Bank, NJ*

INTEGRATED TELECOMMUNICATIONS MANAGEMENT SOLUTIONS

Graham Chen
CiTR
Brisbane, Australia

Qinzheng Kong
CiTR
Brisbane, Australia

IEEE Communications Society, *Sponsor*

Salah Aidarous and Thomas Plevyak, *Series Editors*

The Institute of Electrical and Electronics Engineers, Inc., New York

This book and other books may be purchased at a discount
from the publisher when ordered in bulk quantities. Contact:

IEEE Press Marketing
Attn: Special Sales
445 Hoes Lane, P.O. Box 1331
Piscataway, NJ 08855-1331
Fax: +1 732 981 9334

For more information about IEEE Press products,
visit the IEEE Press Home Page: http://www.ieee.org/press

Printed in the United States of America

10 9 8 7 6 5 4 3 2 1

ISBN 0-7803-5353-6
IEEE Order Number PC5795

Library of Congress Cataloging-in-Publication Data

Chen, Graham, 1956–
Integrated telecommunications management solutions / Graham Chen, Qinzheng Kong.
p. cm. — (IEEE Press series on network management)
Includes bibliographical references and index.
ISBN 0-7803-5353-6
1. TMN (Telecommunication system) 2. Telecommunication systems — Management
3. Business enterprises — Computer networks. I. Kong, Qinzheng. 1957- II. Title III. Series.
TK5105.45.C454 1999 99-047335
658.0546—dc21 CIP

To
Alison and Michael

Contents

Preface

Global deregulation and competition within the telecommunications industry have fundamentally changed the way in which the telco companies and service providers operate. To ensure their future and continuing success, the telecommunications service providers face a challenge to meet the demands for:

- creating the global, ubiquitous communication networks
- improving customer services and maintaining customer intimacy
- creating, deploying, and managing new value-added services

These new market requirements and challenges have forced the restructuring of the telecommunications industry. Traditional telecommunications network service providers focus on providing a ubiquitous network that supports global end-to-end network communication services. The new breed of value-added service providers focus on delivering services to the customers' premises (the last mile of the value chain), and the creation of new services based on the global ubiquitous network.

There is an obvious need for a global ubiquitous network on which end-to-end services can be delivered to the end customers. The technology base of the telecommunications management systems is also changing rapidly. The large number of emerging network and management technologies has greatly enhanced the technology base, but it has also made the global end-to-end solutions more difficult to find. The complexity of today's network and management technology made it more attractive to have a single, unified management framework and technology to support global network operations.

As in many other industries, the constant search for a panacea prevails in the course of developing management solutions. Whether it is a protocol, a management framework, or a management technology, many were introduced to the industry as panaceas. The recent ascendant of IP (Internet Protocol) as the interface protocol to the ubiquitous network is a typical example. Whether IP is the right technology to fulfill this role is immaterial; the intention of the industry is clear that a holy grail must be found to solve all our problems.

Unfortunately, there is no holy grail. The industry has hoped and has been disappointed many times. There is much evidence that the emergence of new technologies, protocols, and frameworks really cannot be controlled. There are already too many of them, and it is certain that more and more technologies, products, and computational architectures will emerge. Without a strong business-solution focused software development discipline, there is a real danger that we will not be able to control the disruptive nature of these new technologies and business process reengineering will turn into expensive technology R&D exercises.

This book discusses some principles and experiences of integrated management. First, we focus on the goal of software development, rather than the standards development. We discuss what the telco business process reengineering really means in the software development sense. The business process reengineering is a paradigm shift. It requires us to fundamentally reexamine the way we conduct our business, define our process, and interact with our customers. However, it also requires us to examine the way we develop the software to support our business, the way we deal with long-term and short-term objectives, technology innovation, and migration.

Second, although a holy grail technology, protocol, or framework to solve our problems seems unrealistic, a well-defined technology management framework is important for the business solution developers. This framework enables us to manage the constant technology evolution and the software solution migration, to define a business-oriented approach, and to model our business independent from the underlying support technology.

This book is organized as follows. In Chapter 1, we discuss the fundamental challenges to the software development of telecommunications management solutions. We introduce the theme of the software life-cycle management and its pivotal role in the business processing reengineering. Chapters 2 and 3 introduce the concepts of TMN, not from a standards point of view but from the view of management software development. Chapter 4 discusses the business and technical requirements for telco business process reengineering and the need for new IT technologies to enhance the functionality of TMN. One of the most important technologies to meeting new business challenges, distributed object-oriented technology—in particular CORBA—is becoming an important part of IT strategy in the telco industry. Chapter 5 focuses on the principles and processes of developing and migrating management solutions to the new distributed OO paradigm, and Chapter 6 deals with the practical technical issues of using CORBA for TMN management solutions.

With the introduction of mainstream IT technologies in the telco industry, the management of multiple technologies becomes a very important issue in the software development exercises. The rest of the chapters cover the issue of developing integrated management solutions. Chapter 7 analyzes the requirements for the technology integration, the purpose of such integration, and the commercial reality of the open standards in this multiple technology era. Chapters 8, 9, and 10 deal with three important aspects of the integrated solutions: the technology integration approaches, the generic objects and application modeling, and the integrated application development architecture. Chapter 11 discusses the software development and technology support for such integrated architecture, and the technology choices to convert it to management software. Finally, Chapter 12 summarizes the theme of software solution life-cycle management and presents further challenges to the software development practitioners in the areas of customer care and innovative services.

We did not write a book about TMN standards, although we heavily referenced the TMN reference model. Our view of TMN is that it is more of a management framework, which defines and partitions different management domains and management functions. The purpose of the functional demarcation is to provide the software industry with an environment to produce reusable components. We focus on the processes and experiences of turning the TMN principles into software components and use the TMN framework as an integration environment to turn components into solutions. Although TMN has defined some specific protocols and computational architecture to implement some management functions, we do not see these protocols and architecture as the only means to implement TMN concepts and principles. A more important task is how to manage the technical evolution of the TMN principles and concepts, rather than implementing TMN concepts using specific technologies.

The main purpose of this book is not to demonstrate how to implement management systems using a specific technology. Although CORBA and CMIP and their integration strategies are heavily referenced, the real purpose is to demonstrate some principles and processes for an integrated software development practice. In recognizing the need for continuous migration and improvement when the technologies and tools improve, these software development practices help us to de-focus our attention from the underlying technologies and refocus to the business requirements for integrated solutions. This is to say that although we cannot control the process and the outcome of the technology wars, we can change the rules of the game to de-focus the importance of the wars.

Graham Chen and Qinzheng Kong
Brisbane, Australia
g.chen@citr.com.au, q.kong@citr.com.au

Acknowledgments

There are many people who have helped, inspired, or encouraged us to complete this book. First, we would like to thank Salah Aidarous for his encouragement, which without we would have never started this project.

We would like to thank Ian Rose for many useful discussions. His insights in many areas, particularly in software development principles and practices, have inspired and enriched a number of topics in this book.

Many people have also provided the inspiration and encouragement which were important for us to complete this project. In particular, we would like to thank Larry Bernstein and Makoto Yoshida.

CiTR has kindly permitted us to use some of their white papers, technical papers, and other publications. We would like to thank CiTR's support and, in particular, Dr. Robert Hendy's support and encouragement.

We would also like to thank many of our former coauthors and research collaborators with who we enjoyed working and have benefitted from through the collaboration and cooperation in many different research areas. In particular, we thank Paul Foster, Ashley Beitz, Jason Etheridge, Jeremy Rixon, Gen Watanabe, Minoru Tanaka, Pierre-Yves Benain, Rubina Hussain, Michael Neville, and Norm Lawler.

Beth Adams reviewed our reference to TM Forum's Telecom Operations Map. We appreciate her kind help.

The IEEE Press editors, Karen Hawkins and Linda Matarazzo, have always provided the encouragement and assistance which helped us to plan and complete the project.

Finally, we thank our parents and children for their continuous support and encouragement. Without their patience and understanding while we spent many weekends and evenings in libraries, we would have never completed this book.

Graham Chen and Qinzheng Chen
CiTR
Brisbane, Australia

1

Challenge of Integrated Business Management Solutions

The global trend of deregulation and competition in the telecommunication industry has increased the urgency for the industry to reexamine the fundamental business model and technology framework. The requirements for new services to be offered at lower cost are forcing telcos to streamline the business processes and optimize technology strategy. These requirements have resulted in the telco business process reengineering (BPR). The software development suppliers, as one of the key players of the telco BPR, are facing increased challenges to provide a technology framework to deliver solutions to meet today's new business requirements and objectives.

1.1 CHALLENGE TO TELCO SOFTWARE DEVELOPMENT

The challenges facing today's software practitioners include:

- The understanding of the requirements for telco business process reengineering
- The multiple technology environment
- The technology integration
- The control of the process

First, we need to understand the fundamental driving forces of the business process reengineering. The global deregulation and competition has forced every telco to fundamentally rethink the way they do their business and particularly the way they deal with their customers. Many newcomers into the telco market operate in a much leaner and efficient manner, they use more off-the-shelf IT technology and products to deliver more customer management functions as a means to gain market share. Retail customers require more sophisticated and cheaper service and product offerings. Wholesale customers require much easier access to the existing telecommunication infrastructure and customer network management functions. The business customers require easier electronic access to their service information and they require more value-add from the service providers to increase the value of their own product and service offers (e.g., e-commerce) to their customers. These driving forces have challenged the industry to rethink the functionality of the management solutions and the software development processes.

Second, the emergence of multiple network technologies and IT technologies has created a much diversified software development environment. We are forced to deal with the reality that we have many different ways to do the same tasks and indeed people have built many systems and components using different technologies. We have different frameworks to design the systems, different tools to build software, and different systems to interact. We have to live with different levels of maturity of the technology we use to build solutions. Under these circumstances, the temptation of *playing* with newest technology and search for perfect solutions is so strong, we sometimes forget the business problems we are supposedly to solve. The business desire to avoid expensive process reengineering and the urge to reengineer technical solutions is sometimes so hard to reconcile.

Third, the consequences of the multiple technology environment have created the challenge to build integrated solutions to business problems. In addition to the technology contributions the IT industry has brought to the telco industry, it also opened a Pandora's box and brought a curse to this industry by propagating the technology wars into the telco management solutions domain. The emergence of IT technologies is very random and the amount of time required for their maturity to support commercial operations is uncontrollable. This sense of not being in control is not helping people who are charged with responsibility to build long-term business solutions, as these responsibilities require a planned release of functionality and orderly enhancement and migration strategy. Whereas many of us may have accepted the notion of the right technology for right problem domains, it is still too hard to practice it in the real world.

Fourth, no software development management process exists to support business process reengineering. On the one hand we curse the existence of legacy systems as the integration with them is so difficult; on the other hand, we really have no processes in place to guarantee that the software we develop today will not be legacy systems tomorrow. There are processes to manage the individual projects which build pieces of the business solutions. Yet which direction these collection of projects will lead us to is unclear and random. The real challenge to the software practitioners is to manage the business solution development life cycle. Although we cannot dictate the solution architecture for tomorrow's business problems, we have enough processes in place to anticipate the future need and to make tomorrow's software development tasks easier.

Finally, the real challenge facing the software professionals in the time of business process reengineering is not to have to do the next business process reengineering. Instead, the solution development for the business problems is implemented as a constant evolution and process improvement.

The foremost question we ask is then, what are the most appropriate technology frameworks in which we can develop applications? What are the frameworks we can use to implement our long-term business objectives and what are the processes leading us to the management software solutions?

1.2 TMN DEVELOPMENT

One of the main initiatives of the BPR in the management solutions arena is the TMN development. The Telecommunications Management Network (TMN) is the most comprehensive framework to manage all aspects of telecommunications, from equipment to customers and from network to services. It defines the business objectives of the telecommunication management, and it encompasses a broad range of management functions and operations required to support telecom businesses.

Yet, TMN is not a technology framework. Although it defines certain technology components in the element management area, it does not equate itself to a management technology. Instead, it defines multiple viewpoints of the telecommunication management:

- It defines the business objectives of the management solutions.
- It provides a reference framework for creating more functions and management applications.
- It defines principles of management architecture and solutions.
- It provides partitions of management tasks into multiple domains and multiple functional areas.
- It provides an abstraction view of management process models and information models.
- It defines an instantiated computing architecture to support communications and interoperability at the network and element management level.

A misconception in a segment of the industry is to equate TMN to a technology, specifically, to equate TMN to OSI technology. While OSI is a technology useful to the management of telecommunication equipment, it covers only a small portion of the management framework and provides only one of the functions—to achieve the protocol level interconnection. A large part of TMN framework, particularly in the area of integrated software architecture, has not mandated any specific technology as its choice of implementation. It is up to the technology and solution creators to decide the most appropriate technology strategies to implement TMN.

This is indeed the direction taken by many technology creators and TMN solution providers such as:

- Telecommunication Information Network Architecture Consortium (TINA-C)
- TeleManagement Forum (TMF, formerly Network Management Forum)
- Object Management Group (OMG) Telecom Task Force

TINA-C has defined a Service Architecture [92] and Network Resource Architecture [95] which are instantiations of TMN service and network management functions. These architectures are supported by a Distributed Processing Environment (DPE) [96], a computational environment supporting distributed processing and application development.

TMF defined a Telecom Operations Map (TOM) [99] which instantiates the TMN management functions with a set of well-defined business processes and object models known as the SmartTMN [2]. This map is also supported by a computational environment defined as the Technology Integration Map (TIM) [100]. The telecom operations map provides TMN solution developers with an industry agreement on business processes and information flow-through, and the technology integration map provides them with a guide for management technology procurement, development roadmap, and conformance.

OMG's Telecom Task Force defined a CORBA-based architecture to support OSI-based network management functions [91]. This architecture supports most TMN network management concepts and operations. The task force worked closely with TMF and X/Open Group to define CORBA/OSI interworking gateway to facilitate the introduction of CORBA technology into TMN management domains [102]. OMG's Business Objects work also targets the definitions of business processes and objects for implementing TMN functions.

In addition to reinforcing the view that TMN is a framework of telecommunication management rather than a technology, these initiatives represent challenges to TMN development. These industrial programs have manifested the fact that TMN's future relies on the technology instantiations to produce solution architectures. Otherwise, TMN will always remain a set of principles of telecommunication management, useful only as a reference model for verifying business objectives.

The first technical challenge to TMN is the use of the IT technology to enable the development and deployment of its functions. One of the strong requirements of telco BPR is to define a TMN computational architecture for deploying management functions in a much more distributed and large scale networks. This requirement exists in both network management functions and service management functions. The network management solutions have to be deployed to manage nation-wide or global networks with multiple technologies and ever increasing network resources. The service management solutions have to support a widely distributed, highly dynamic, and increasingly diversified service deployment scenarios. With the focus of TMN management shifting more toward the end customer management, the highly scalable solutions are sought to support on-line customer access to, and service interaction with TMN systems through a very wide range of access mechanisms. There does not exist any defined technology framework in TMN to satisfy these requirements. New technology strategies have to be defined to create these solutions.

The second technical challenge to TMN is to avoid turning BPR into a sequence of expensive R&D projects and TMN development into a technology play. Instead, the TMN development requires a strong business focus and disciplined management process. The introduction of IT technologies into TMN has created almost infinite number of ways to achieve the same results. Many of the industry's TMN development exercises failed because we embark on a search for a perfect solution. In many cases we end up with a sequence of (new) technology evaluations rather than an instantiation of solution suitable for our business objectives. The IT technology and products emerge with a much quicker pace than the business requirements. We must understand that the business solution architecture needs almost certainly to out-live a specific technology, no matter how *hot* the technology.

The third technical challenge is to provide an integrated TMN solution framework. TMN defined an integrated management framework with different viewpoints at different levels of abstraction. This has created difficulty and confusion to the TMN solution development. The vertical management layers were easily and wrongly mapped into totally isolated software architectures and platforms among which no processes and information are flowing through. The conceptual separation of horizontal management functions was interpreted literally to create vertical application stovepipes which are very hard to integrate. With the emergence of more and more network technologies and management technologies, the industry will end up with more specific functional pieces of software (such as SNMP-based element manager for a vendor's ATM switches, or OSI-based network manager for another vendor's SDH networks) but no integration strategy to put them together to form business solutions. Even though some of these components exist in some standard technologies, they still do not equate to business solutions. The reality is that many of them are in proprietary technologies and were never designed to interact with each other! The challenge to the industry is to define integrated architectures so that these components can be used to form business solutions. Based on this architecture, the integrated management applications can be developed and these applications will combine the functions of the network equipment, value-added services, and customer care to deliver the customer services and to achieve the customer satisfaction.

The fourth challenge is how TMN will evolve and mature with new technologies. Telco industry will forever be in a state that business solutions are supported by a large range of technologies with different levels of sophistication and maturity. The development will always face the following issues:

- how to interact with existing legacy systems
- how to deal with multiple new technologies
- how to manage the overall life cycle of the solution development aimed to reduce the chance of building legacy systems

It is the responsibility of the software practitioners of TMN development to manage this evolution process. We need to clearly understand that in this process, we cannot break from the past and nor can we dictate the future. In our solution development, we have to interact with the legacy systems by, say, using object wrappers. Also we have to be prepared that the components we are building now will in the future be replaced or adapted to the future solutions. A good software practice is, for instance, to specify fully the functionality of the components and the external behaviors of the interface, and at the same time, to define the minimum requirements for integration by designing very generic and simple interfaces. The specifications should not only include computational specifications (information and functional specifications and programs) but also include the proper documentations. One of the less attractive attributes of a legacy system is that over time it evolves but the evolutionary interfaces and functionality is not documented well. Therefore it becomes more and more difficult to interface with other newer systems. One of the biggest weakness with business life cycle management is the lack of good (and standard) documentation processes. To develop and manage this process should be treated as the integral part of the TMN solution management.

1.3 TMN STANDARDS DEVELOPMENT

The challenge to TMN development also exists in the standards area. If we examine the TMN standards development so far, we could categories them into:

- *Frameworks and architecture* such as management frameworks and transport architecture.
- *Interoperability* such as management functions (SMF) [77], [78], communication (SMC), and information (SMI)
- *Specification of operations* such as a large quantity of specifications and object models defined by using the *Guidelines for the Definition of Managed Objects* (*GDMO*). These are usually very complex and detailed.

Let's examine how the standards can help us develop TMN solutions for business problems. The first type of standards defines concepts and principles. These are important to guide us to the overall solution architecture and the style of the technology selection, such as client/server technology or distributed object-oriented technology. However, there are no clear definitions of conformance to the overall architecture, nor is it necessary. People are entitled to and will implement TMN principles using different approaches.

The interoperability standards help us to build basic plumbing infrastructure to exchange information. These are so far largely useful at the element level. Since the highest value of the TMN business solutions exists in the high levels (end-to-end network functions, service and customer management, business applications), the technical and business innovation at the protocol level is a nonstrategic issue. It makes sense that we use standard protocol to exchange information at the element level.

The operational standards are problematic. There are a large quantity of TMN standards defining how to *do a thing* in a TMN way such as managing connections, reporting faults, managing lease lines, etc. These standards are overloaded. They intend to guide people to implement specific management functions in commonly accepted ways, i.e., they are object specifications of systems or components. On the other hand, they are always meant to be interface specifications for interoperability purposes. That is, two operation systems implementing the same standard specification are meant to interoperate.

In software engineering practice, these two roles impose hugely different requirements on the way objects are modeled. A specification of a system always requires a detailed object analysis and object definition. A specification for an interoperability purpose is always more effective if it is simple and abstract. The results of many TMN specifications are that they may just be detailed enough as specifications for operational systems, but they are too complex to be interface specifications for application level interoperability.

This leaves a major gap in the standards that there are very few, if any, specifications designed for application integration. People complain about the complexity of TMN standards, because they have to implement the whole complex and detailed specifications to claim conformance—for the purpose of interworking or integration.

In order for TMN to become a management framework for applications to interact, interwork, and to grow in a building block fashion, the standards have to be simplified. The value of a standard specification facilitating interworking between functional components is far greater than a standard defining operational details of a component.

1.4 THE PURPOSE OF THIS BOOK

This book deals with the technical challenges of building integrated TMN solutions in a multiple technology environment. It provides the principles and processes one could adopt to deal with this complex issue. Each chapter deals with a different aspect of this theme.

The integrated TMN solutions have strong and urgent business requirements. These requirements dictate our business objectives. They provide reasons why we need business process reengineering. They provide strong input into the process of defining integrated solution architecture. Clear understanding of these requirements at any stage of the solution development process is crucial for us to stay focused.

The introduction of IT technology into telco industry, in particular, the distributed object-oriented technology has greatly enhanced the strength of the integrated TMN solutions to answer today's business challenges. However, this has also introduced a significant degree of complexity in the technical solutions and the management of TMN development. Before we adopt the distributed technologies, we have to be convinced that strong business requirements do exist for adopting these technologies. We must have a firm understanding of the complexities and difficulties we will face in adopting these technologies.

Migration from centralized development environment to the distributed environment requires a fundamental paradigm shift and an appreciation of a wide range of new design principles and models. To avoid the risks of our solution development efforts turning into expensive R&D exercises, we need to be prepared for this paradigm shift and understand this sometimes painful migration process. We need stay focused to the business requirements and have a clear strategy of dealing with the emergence of new technologies, new standards, and new and somewhat immature products.

The notion of the right technology for the right problem domain is gaining more support in the industry. This reflects the maturity of the industry toward new IT technologies. This notion asserts that it is the business problems and challenges caused by deregulation and competition that drive the progression of the TMN solution development process. Technologies are important but they are only the means to achieve the business objectives. This paradigm shift is crucial as it helps to achieve the following goals:

- TMN development becomes more customer solution focused.
- Development efforts are more result oriented, rather than technology indulgence and experiments.
- Solution development processes are more concerned with end-to-end needs.
- Software practitioners are more concerned with integrating different technologies rather than constantly picking the best technology.

These have helped us focus to develop integrated solutions and avoid technology wars.

The technology integration has presented us with some new challenges. First, we need to understand the purpose of the integration. The ISO and ITU-T standards movement in the past decades has a very strong focus on interoperability. However, the TMN solution development requires more than interoperability. It requires an environment for application interworking and component-based application development. Second, we need to understand different integration strategies for different requirements. Third, we need to define an integration framework which consists of a well-defined business process model and information model, i.e., we need to *know* and be able to *describe* our business better before we build management solutions. To face the challenge from a deregulated market place, we need to define a strong customer-oriented view of TMN. Traditionally, TMN defined service providers' or equipment vendors' view of the telecommunication management. This should be complemented by a customer focused view, centered on the management of customers' service interactions with their service providers.

This book is not about building telco operation systems. It is not about how to build a TMN fault management system or VOD service. It is about establishing management processes and development principles that will lead to an integrated environment for new and existing operation systems.

This book is not about building CORBA or OSI solutions. Although CORBA is used throughout the book as an example of distributed object-oriented technology, and it is indeed becoming the strong technology choice for the telco industry among others, very few principles and processes discussed in this book are dependent on the choice of CORBA as the underlying technology. Indeed all the principles and processes discussed in this book are about how to build long-term business solutions that out-live the individual technologies used in the solutions. We will demonstrate that the rapid emergence of IT technologies and their adoption in the telco industry does provide a serious challenge to the life cycle management of our solutions. The divergence of technology is now so strong that every functions can be supported by a large number (and still growing) of technologies and products. This reality simply means that the notion of best technology does not exist. We cannot advise anyone (we doubt if anyone can) what is the best technology for the future of TMN development, and it is very irresponsible for us to do so. However, the principles and processes do exist which can help one manage the process of solution development, technology selection and migration. This will be the focus of the book, and this will be the area to which we hopefully add value.

2

TMN Overview

From its initial operation, the global telecommunication network required management capabilities. Whether this management capability lives within the communication network and its elements (devices), or it lives in a separate software entity, there must be some management functions which ensure that the global telecommunication network is operating smoothly. This management capability includes:

- All communication equipment is operating smoothly. The equipment can perform its function with a reasonable and measurable quality. Any errors and faulty behavior can be detected (either automatically or manually) and repaired. The operation of these elements can be monitored and statistics of various kinds can be gathered and reported to operators or central management systems.
- All communication equipment in the telecommunication network supports some *network* functions. The elements can be configured to support data transmission channels. These channels allow the information to be switched and transmitted from one part of the network to any other part of the network. Again, the management capabilities are required to ensure that this information transmission is conducted in a controlled way, with a measurable quality, error detection and repair capability. Also the statistics of various kind relating to the information transfer need to be collected, correlated, and reported to various entities such as operation control, monitoring, and administration systems.
- The telecommunication network must be able to provide services to the end customers, such as telephone services or information services. These services are supported by the networks and their elements. Again, some operation systems are required to support different aspects of provision of services. The services have to be delivered and activated; any service faults and customer problems have to be detected, located, and resolved; the services have to be operating with a reasonable and measurable quality and the services have to be billable and paid for by the customers; therefore the accounting systems have to be deployed to generate bills for the network resource usage and any value-added service components.

These capabilities directly impact the success of telco's business: to provide telecommunication services to the users. The business processes of various kind, and with a different degree of automation, were created to provide these capabilities.

Two major requirements for management systems emerge from the evolution of the telecommunication network which led to the creation of the telecommunication network management framework. The first one is the introduction of computer technology to increase the level of automation in these business processes. The second one is the need for different portions of the network and different kinds of devices to interconnect with each other in a common way. Without meeting these two requirements, the growth of the telecommunication network is limited since this growth is fundamentally based on the heterogeneity and scalability of the network and services deployment.

These requirements led to major international standardization initiatives in the telecommunication and computer industry to define and standardize management frameworks which support the automation of operational systems and the interconnection between telecommunication and computer networks. These frameworks encompass the following elements crucial to provide management capabilities:

- A reference model or reference architecture in which the management capabilities are defined and scoped. This allows the boundary of the management capability of the overall systems architecture to be defined.
- A functional model describing various functions the management framework has to support. This partitions the management tasks into a set of interrelated functional components. This model allows the software architecture to be derived and implemented to support various management capabilities in a standard way.
- An information model in which all management domains and tasks can be described or *specified.* This model supports the interconnection and interworking between different networks, elements, and management functions. It allows the functional components to operate in multivendor and multitechnology networks since the *contents* of the various components of the network are specified using the standard methods and techniques.
- A computation model which includes communication, operation, and protocol definitions. This model supports the computational aspect of the automation and interconnection.

TMN is a product of this international standards movement. It is inspired by the fundamental business requirements to provide a management framework in which the automated management capabilities can be deployed to manage the interconnected, heterogeneous, and large-scale telecommunication networks.

2.1 TMN INTRODUCTION

TMN (Telecommunications Management Network) is a telecommunication management framework that supports the management requirements to plan, provision, install, maintain, operate, and administer telecommunication networks and services. The concept of TMN was first proposed in 1986 and formally defined in 1988 by ITU-T (formerly CCITT). In 1989, the *Principles for TMN* was published as an ITU-T Recommendation M.3010 [82]. Following that, an initial set of implementable TMN interface recommendations were published. These include:

- Generic Network Management Information Model (M.3100 [83]),
- Lower Layer Protocol Profiles for the Q3 Interface (Q.811), and
- Upper layer Protocol Profiles for the Q3 Interface (Q.812).

TMN is conceptually a virtual network that is built over a telecommunication network. This management network interfaces with the telecommunication network resources at several different points to exchange management information with the network. A TMN management system may use parts of the telecommunication network to provide communications. The relationship between TMN and the telecommunication network is illustrated in Figure 2.1.

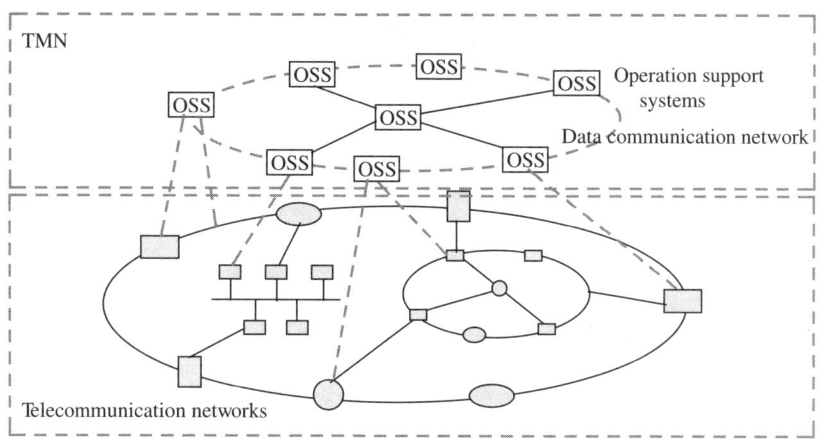

Figure 2.1 TMN and telecommunication network.

Figure 2.1 shows the components of TMN including the data communication network interconnecting various TMN operation support systems, each providing different management and operational support functions. The data communication network can be a separate network (such as computer network) for management purposes, or it can be part of the communication network which allows the transmission of management data. Some of these operation systems performing different functions can reside in the network or its elements.

The functionality of the TMN consists of mainly the abilities to exchange management information between different components. This information exchange can take place at different locations within a TMN environment using different formats. It can also take place across the boundary between a TMN environment and another TMN environment or a non–TMN telecommunications environment. The functionality also consists of the abilities to handle management information, such as to analyze, manipulate, and deliver management information.

The importance of TMN was recognized by many regional and international standards bodies, such as the European Telecommunication Standards Institute (ETSI), TMF, TINA-C, etc. TMN is seen globally as the basis for the management of today's telecommunications networks.

As a management framework, TMN provides an environment for interfacing computer systems with telecommunication network resources to provide various management functions at different levels. The framework allows the management of business information between different components (operation systems, communication equipment, network resources, and computer systems) and provides control over service operations and network functions.

Work describing TMN framework has been widely reported in literature. In this chapter, we focus only on the following aspects of the TMN:

- *TMN architecture*—an architectural definition of TMN entities, their functional roles in the framework, the communication interfaces between them, and their interrelationships.
- *TMN management layers*—the management functionality for different layers of a telecommunication network.
- *TMN management functional areas*—the management functionality to support the operation, administration, maintenance, and provisioning requirements of a telecommunication network.
- *TMN management platform*—the communication protocol to transport management information between different TMN entities and management services.
- *Management information models*—the methodology of information modeling and the abstract view of the telecommunication network resources to be managed in the form of managed objects.
- *TMN management applications*—the development of TMN-based telecommunication management applications.

2.2 TMN ARCHITECTURE

The TMN architecture [82] is defined from three perspectives and three architecture models are defined respectively. These architecture models are:

- *Functional architecture model*—it defines the functional components of TMN and the interface between these components.
- *Physical architecture model*—it identifies the physical components of TMN and connectivity between these components.
- *Information architecture model*—it provides an object-oriented methodology to model telecommunication networks and services and the management functions.

2.2.1 Functional Architecture Model

The TMN functional architecture describes the different functional components. A typical TMN system will consist of a set of interrelated functional components. The interactions between these components are defined as reference points. They dictate the information being communicated between functional components when these components are distributed within a TMN environment.

TMN functional architecture also defines the interactions with other external actors including:

- TMN management information users (customers)
- Peer TMN conformant systems
- Non–TMN conformant systems

Figure 2.2 depicts the TMN functional architecture with five defined TMN functional components and their relationships. These functional components are:

- *Workstation Function* (*WSF*)—This function supports the interactions with TMN management information users. This function is partially in a TMN environment. In

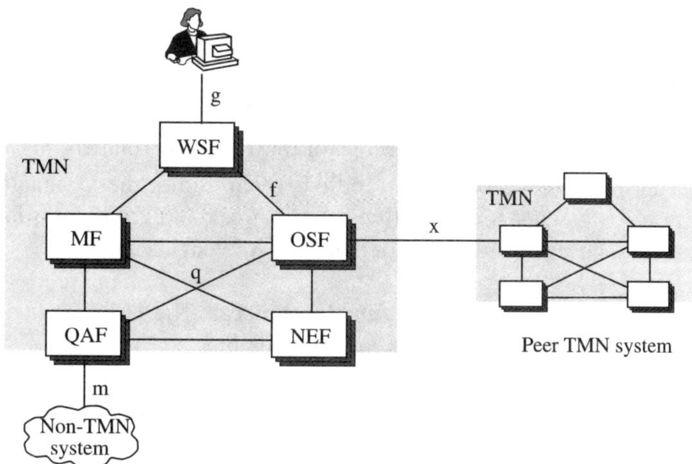

Figure 2.2 The TMN functional architecture.

a typical TMN implementation, this function is acting as a mediation device between TMN management functions and user applications.

- *Operation Systems Function* (*OSF*)—This function supports processing of management information to monitor, coordinate, and control telecommunication operations (including TMN management operations). This component embodies the entire TMN management functions a user can access.
- *Network Element Function* (*NEF*)—This functional component represents the management capabilities supported by the network elements. The NEF provides the network element level support for OSF.
- *Mediation Function* (*MF*)—This functional component facilitates the information passing between OSF and NEF or QAF when different internal q reference points are used by these components. It provides support to store, adapt, filter, threshold, and condense information when facilitating interactions between different q references.
- *Q Adapter Function* (*QAF*)—If part of the network or an element support TMN functions, then OSF uses NEF to access its functionality. But if they are non–TMN networks or elements, the OSF needs to access them through an adapter which provides the protocol and information translation between a non–TMN and TMN environment. This translation is provided by the QAF.

Various reference points are specified in order to define the service boundary and interactions between the TMN functional components as well as their interactions with the external components:

- The interactions between the internal TMN functional components are conducted by using q reference points (further divided into q3 and qx). An exception is their interaction with WSF, where f reference point is used.
- The provision of different q reference points (q3 and qx) allows interactions between different internal TMN components to have different behaviors. If two components interact with two different q reference points, then a mediation function is required to facilitate the interaction. The mediation function translates two different q reference points to achieve the interactions. The mediation function is not required if the same q reference point is used. The most commonly defined q reference point is the q3 reference point.

- The OSF component uses x reference point to interact with its peer in other TMN systems to exchange operation systems functions.
- TMN user applications use g reference point to access WSF in order to gain the access to TMN management functions. This reference point is identified in the architecture but it is outside the scope of the TMN environment.
- The m reference point is specified which allows the Q adapter to interact with a non–TMN management entities. Again, this reference point is identified in the architecture but is outside the scope of the TMN environment.

In addition to the above listed functional blocks, The TMN architecture also identifies a set of components as the basic functional building blocks. Examples of the functional components include: management application function (MAF), workstation support function (WSSF), information conversion function (ICF), directory system function (DSF), etc. Each functional block contains a various number of functional components. A data communication function (DCF) is also identified as a functional component, however, it is located between different functional blocks, rather than contained in any of the functional blocks as their component. A functional block can use DCF to exchange information with another functional block. If different subnetworks are interconnected, the interworking functions will be part of the DCF.

2.2.2 Physical Architecture Model

The purpose of a physical TMN architecture model is to identify and correlate the physical entities to functional components. These physical entities can be hardware or software modules which support the functionality specified in the TMN functional architecture model. This TMN physical architecture model is depicted in Figure 2.3.

The components of the physical architecture include:

- *Data Communication Network (DCN)*—This is the physical communication network within TMN which supports the communication requirements between different components of TMN and the distributed deployment of a TMN system. It can be a separate physical computer network or part of the telecommunication network it manages.
- *Network Elements (NEs)*—Network elements are telecommunication equipment and support equipment in the TMN environment. The network element function, NEF, is normally supported by the physical network element conforming to TMN.

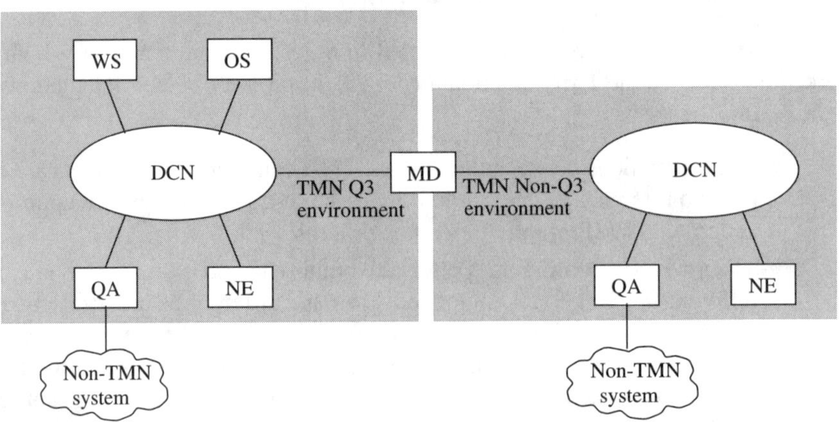

Figure 2.3 The TMN physical architecture.

Otherwise, a non–TMN conformant network element needs to be integrated into a TMN environment by interacting with a QAF.

- *Mediation Device* (*MD*)—A mediation device is required between two internal TMN components if different Q interfaces are used. The mediation device will provide mediation functions to allow the information translation between these interfaces.
- *Q Adapter* (*QA*)—This is a device that allows a non–TMN component (a device, a network or a non-TMN management system) to be managed by TMN components. It supports the Q adapter functions. In many cases, it is a software implementation which forms part of the operation system functions interacting with the non–TMN environment.
- *Operations System* (*OS*)—An operation system is a software used to monitor, coordinate, and control telecommunication operations and management functions.
- *Workstations* (*WS*)—Workstations are used by operators to access the TMN management environment, and they support WSFs.

To support the communication between different TMN components, the TMN physical architecture defines a number of interfaces. These interfaces apply to their corresponding reference points in the functional architecture.

- Q interface is applied at q reference points.
- F interface is applied at f reference points.
- X interface is applied at x reference points.

2.2.3 Information Architecture Model

The TMN information architecture is based on an object-oriented paradigm. The OSI system management principles are fully adopted by the TMN information architecture. The major characteristics of this architecture include:

- All physical and logical network resources to be managed are modeled as *managed objects.* A managed object is an abstraction of a network resource, a relationship between resources, a combination of resources, or a logical resource.
- Each managed object has a set of *attributes* which are visible to a management system.
- Each managed object is the target of some *management operations.*
- Each managed object may initiate *notifications.*
- Managed objects that share the same attributes, operations, notifications, and behavior are grouped into *object classes.*
- The *object class hierarchy* is established by object inheritance and multiple inheritance is supported.
- Managed object instances are maintained in a *management information base* (*MIB*).
- Each object instance in the MIB is identified by its object class identifier and its *distinguished name.*
- GDMO (Guidelines for the Definition of Managed Objects, ISO 10165-4, ITU-T Recommendation X.722), and ASN.1 (Abstract Syntax Notation 1, ISO 8824, ITU-T Recommendation X.208) are used for objects and attributes definitions.

The TMN information model is depicted in Figure 2.4. The TMN information archi-

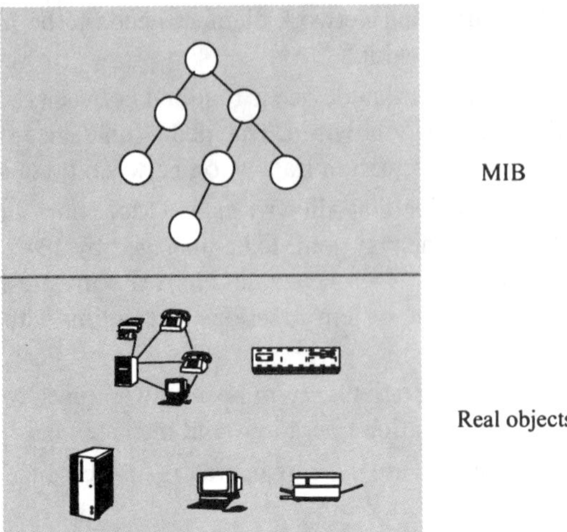

Figure 2.4 TMN information model.

tecture also fully adopts the OSI manager/agent concepts and the relationship model between manager, agent, and objects. These will be discussed in more detail in the next chapter.

2.3 TMN SYSTEMS

The three TMN architectures define three different views of a TMN system. The functional model defines what the TMN system does, the physical model defines what components it consists of, and the information model defines how a TMN system is described.

A TMN management system is a set of management software components cooperatively performing management functions. The TMN system manages telecommunication networks and all services supported by the networks. The TMN architectures are abstract representations of these different software components comprising an application.

DCN functions are normally supported by a computer network, and this network facilitates the open communication between TMN components and with other TMN systems. As Q3 interface is the most common interface between TMN components and TMN systems, DCN's main function is to support Q3 communication.

TMN workstation functions model operator's behavior and operator's access to TMN functions. They are supported by management applications.

TMN management applications consist of operation systems which manage network resources. In the TMN model, the telecommunication network resources are either TMN conformant resources or proprietary resources.

The TMN conformant network resources are modeled as managed objects by using the modeling tools in the information model. This modeling includes the managed objects, attributes, operations, and relationships, in particular the rules defining the managed information base. The TMN-managed objects present an abstract view of the managed physical resources. These managed objects may support Q interfaces, which enable the management information on managed objects to be passed between different components or TMN systems.

If different Q interfaces are supported by different components, then a mediation device is required to provide mediation functions. The mediation function translates between different types of Q interfaces. If every component supports Q3 interface, then the mediation function will not be required.

If the network resources support only the proprietary interface, then a Q adapter is required to provide Q adapter function so that the Q3 interface can be used to communicate with that system.

A TMN operations system defines part of the overall system semantics. It is normally developed as a component of the management application supporting a particular function, such as managing billing information. This application component may not interact with any part of the physical network. In a typical TMN system, multiple operation systems can be developed.

Alternatively, an operation system can be developed as a management entity which interacts with the physical network to manage and control its behavior. These management functions are provided by controlling and manipulating the managed objects (directly, through mediation devices or through Q adapters). These management entities support other operation systems which act as part of management applications. They communicate using Q3 interfaces.

Figure 2.5 depicts a typical software system implementing TMN architecture. The grayed area defines the boundary of the TMN system. The system contains a communication platform which offers DCN functionality by supporting Q3 communications. This platform

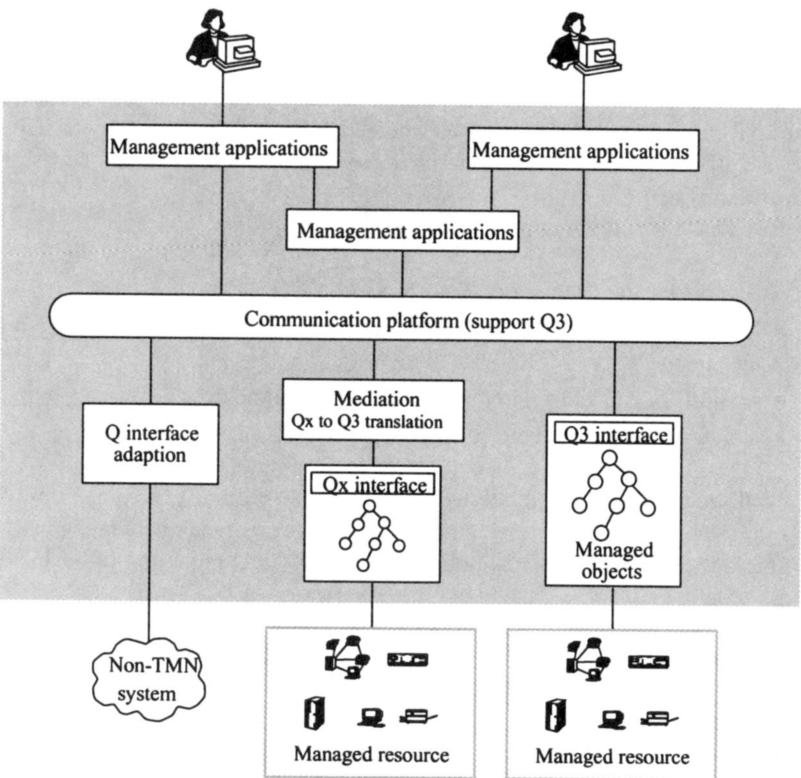

Figure 2.5 TMN management system.

facilitates open peer-to-peer communication between various systems components. Six operations systems are depicted in this software architecture; three of them are managing network resources. One manages a portion of network resources supporting Q3 interface, another manages resources supporting Qx interface so a mediation device is developed to conduct the interface translation. Finally, the third operation system manages a non–TMN environment, and it employs a Q adapter software component to achieve interface translation. Three operations systems are providing management application functions and supporting operator's access to the TMN system.

2.4 TMN MANAGEMENT LAYERS

When TMN management architecture is used to provide the management capabilities for telecommunication business activities, it is necessary to characterize the management environment into different domains with different management focuses and requirements. This characterization has resulted in the partition of management environment into multiple layers, each with different management objectives and technology requirements.

The partition of the management environment into multiple layers also defines the business value chain from the management of telecom business, to the management of products and services to the customers, to the management of networks and finally to the management of the telecommunication equipment. Each layer at the lower level of the value chain provides functional support for their previous level and each upper layer represents more value-add to the final business objectives.

2.4.1 Business Management Layer

The business management layer (BML) has responsibility for the total enterprise, and BML is the layer at which agreements between operators are made.

This layer normally carries out goal setting tasks rather than goal achievement but can become a focal point for action in cases where executive action is called for. This layer is part of the overall management of the enterprise, and many interactions are necessary with other management systems.

BML has the following principle roles:

- support for the decision-making process
- support for administration, operation, and management (AO&M)-related budget management
- support for AO&M-related work force management
- maintain aggregated enterprise data.

BML does not necessarily correspond to any specific software solutions. Although it is a focus of the business process reengineering, its main software activities involve the modeling of the business environments, policies, and high-level modeling of business objects and business processes.

2.4.2 Service Management Layer

The service management layer (SML) is concerned with, and responsible for the contractual aspects of services that are being provided to customers or available to potential new customers. Its principle roles are:

- to control and manage interfaces to telecom service customers
- to manage interactions with service providers
- to manage interactions between services
- to maintain quality of service data.

Service management functions are usually implemented by the operation support systems which form the core part of the TMN solutions. They interwork with the network management and element management components to provide the support for telecommunication services to end customers.

2.4.3 Network Management Layer

The network management layer (NML) is responsible for the management of a network with the support of the element management layer (EML). The NML is responsible for providing a technology independent view of the network resources and functions to service the management layer. NML has four principle roles:

- to control and coordinate the network view of all network elements within its scope or domain
- to provide, cease, or modify network capabilities for the support of service to customers
- to maintain network capabilities
- to maintain network data and interact with the service management layer on performance, usage, availability, etc.

In the typical TMN solutions, software is developed to manage the networks and provide a NML functionality. This functionality usually presents an abstract network view of the physical telecommunication network resources.

2.4.4 Element Management Layer

The element management layer (EML) manages each network element on an individual or group basis and supports an abstraction of the functions provided by the network elements. Network elements include many types of telecommunication equipment and associated support equipment and systems. Examples of network elements are transmission systems, switching systems, multiplexers, signalling terminals, front-end processes, computers, cluster controllers, file servers, etc. All of them can be represented by a set of managed objects and managed by EML. EML is responsible for providing a vendor independent view of network elements to network management layer.

The principle roles of EML are:

- to control and coordinate a subset of network elements either individually or collectively
- to maintain statistics, logs, and other data about elements

In a typical TMN–based solution, software components are developed to provide the EML functions.

2.4.5 TMN Layers and Its Value Chain

The partitioning of TMN management layers is only an exercise to divide the whole management tasks into smaller and logical chunks. It provides an environment to view the whole management tasks from different viewpoints. In particular, it defines the value chain from the customers to the network elements. At every interaction point between two different layers, it maps the business objectives of the upper layer to the supporting functions and services at the lower layer. Figure 2.6 illustrates the value chain of TMN model.

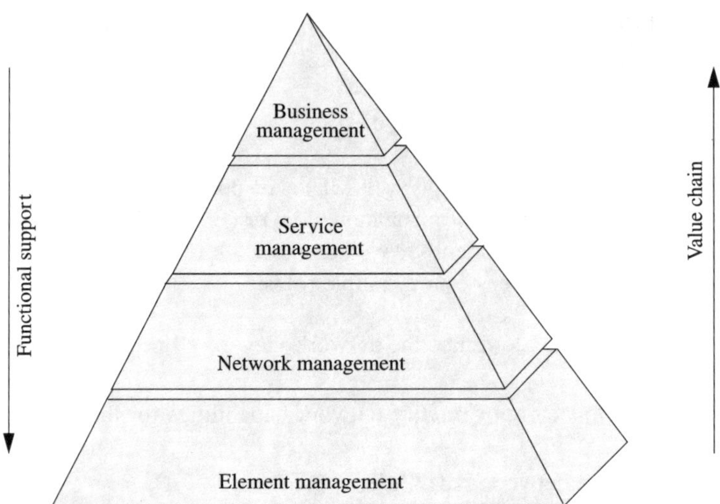

Figure 2.6 TMN functional layers and value chain.

This layered model does not suggest a multilayered software development requirement for implementing TMN framework.

2.5 TMN MANAGEMENT FUNCTIONS

A TMN management function is a cooperative interaction between application processes for the management of telecommunications resources. The application processes supporting TMN management functions form the core of both managing systems and managed systems.

The objectives of the TMN is to provide a framework for telecommunication management. Its architecture defines five management functions:

- *Fault management*—it encompasses fault detection, isolation, and correction of abnormal operation or improper behavior of resources.
- *Configuration management*—it controls the life cycle of the interconnection services in order to provide for continuous operation. The life cycle of the interconnection services include the preparation, initialization, operation, and termination.
- *Account management*—it enables charges to be established for the use of resources and for costs to be identified for the use of those resources.
- *Performance management*—it supports the functionality required to monitor and report on the behavior of resources and on the effectiveness of the different activities that can be achieved through the use of the resources.
- *Security management*—it supports the application of security policies by the creation, deletion, and control of security services, the distribution of security-related information and the reporting of security-related events.

Each management function has different semantics in different layers. The relationships of the five TMN management functions (known as FCAPS functions) and the abstract management layers illustrated in Figure 2.7.

Each FCAPS function at different management layers has different meanings. Using network management (including network element management) layer and service management layer as an example, the management functions in service management have a different

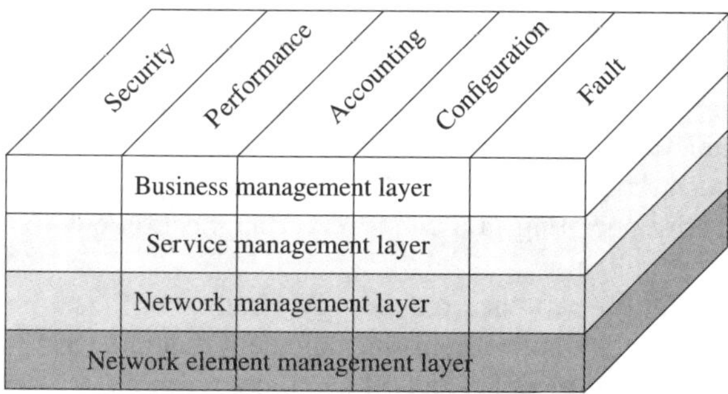

Figure 2.7 TMN management layers and management functions.

orientation and broad scope. The concept of resources and managed environment is no longer limited to the network physical resources. They include systems, services, business objects, and customers. All five TMN management functions have extended meanings in the service management layer compared with the ones in the network management layer. Table 2.1 presents the differences in management functions between the network management layer and the service management layer.

TABLE 2.1 TMN Network and Service Management Functions

Categories	Network Management	Service Management
Scope	Network and Network Elements	Services, Customer profile associated with services
Fault management	Alarm surveillance Fault localization & correction Testing Trouble reporting and tracking	Detection, diagnosis, and correction of abnormal service operations Preventive procedures Trouble ticketing
Configuration management	Connection management Resource configuration management	Service life cycle management Customer life cycle management
Account management	Metering (call event record, CER) Charging (CER → Type, Units)	Billing (units → customer billing)
Performance management	Traffic status, traffic performance monitoring	Maintaining the agreed QoS

The fault management, for example, has extended its scope from managing faults in network and its elements to managing the faults in services. It is more concerned with the restoration of a service to its subscribers within a time frame set by a contractual agreement.

In the area of performance management, the service management is concerned with maintaining an agreed level of quality of service. It will not only make sure that the network and its elements are performing satisfactorily, but also make sure that the service offered to its subscribers is performing satisfactorily. The management function will prevent any possible degradation of quality of services caused not only by the network performance, but also by the environment the service is offered. For example, a less scalable service deployment system may perform badly when more customers subscribe the service, even if the network is performing well.

The accounting function in the service management layer is very different from that of network management. The management function is concerned with the implementation of tariff policy and mapping accounting information in the network management layer (which is primarily concerned with the usage of network) into the billing agreement between the service providers and service users.

2.6 TMN TECHNOLOGY FRAMEWORK

The technology framework of TMN is an object-oriented, open systems architecture. The technology framework is largely designed and deployed for the element management and network management layers. It views a TMN network management system as a networked, interconnected entity operating in a client/server application environment with a peer-to-peer communication paradigm.

The foundation of the technology framework is the OSI (Open Systems Interconnection) network and systems management function (SMF) and its systems management information (SMI) model. The OSI technology framework defines the application entities, their communication paradigm, the operations and services they provide, and the application development and deployment environment. The technology also provides an object-oriented information modeling framework to describe the entities in the managed environment, management entities, and their relationships.

2.6.1 Management Information Modeling

An important element of TMN framework is the management information model. This information model describes the relationship between the real network to be managed and the management environment. The object-oriented technique is used in the information modeling. In the TMN environment, each managed resource is defined as a managed object. This information model defines:

- what managed objects are
- what they are composed of
- what they can do
- what can be done to them
- how they relate to each other
- how they are named

A managed object is an abstraction of a physical or logical resource. It is defined for the purpose of managing the resource. An object has a set of attributes which describe the characteristics of the object.

An object may have the capability of issuing notifications and is the target of management operations. The behavior of an object can also be modeled to reflect the management view of the underlying resource and to define the effect and constraints of operations on the object.

In order to define managed objects in a standard way, an abstract object modeling language, GDMO, is used. GDMO provides detailed specification of how the managed objects should be modeled. Another set of abstract notation, referred to as the Abstract Syntax Notation One (ASN.1), is also used. It allows the more detailed specification of the data types of the attributes in each object.

Managed objects that share the same attributes, operations, notifications, and behavior are grouped into object classes. One object class can be derived from another object class. The original class is referred to as the superclass and the derived class is referred to as a subclass (of the super object class). All characteristics of the superclass is inherited by the subclass. The Object Class Hierarchy is established by the inheritance. In OSI–based TMN environment, multiple inheritance is supported.

2.6.2 TMN Network Management Function

A TMN management function is essentially a cooperative activity involving two application processes, one in a managing system sending management requests and the other in a managed system performing management operations according to the requests.

In order to perform these cooperative activities, some agreements are needed between peer communication entities:

- *A communication agreement*—This includes communication protocol and a set of generic management services. This agreement establishes the basic understanding between management processes involved in the communication.
- *A functionality agreement*—This includes a set of specific management functions such as:

  —Object management function
  —State management function
  —Relationship management function
  —Alarm reporting function
  —Event report management function
  —Log control function
  —Security alarm reporting function.

These functions are the building blocks needed to construct management entities.

- A set of object definitions accepted by all management entities as the logical representation of the managed environment. All management activities are carried out by performing certain management operations on a set of managed objects; or emitting notifications on the status change of managed objects. These objects must be represented in a common form which is accepted by all management entities.

A management platform is a computer software architecture and infrastructure which support the functionality required to carry out the management activities. Several aspects of such a platform are further discussed in the following sections.

2.6.3 OSI System Management Protocols and Services

Due to the similarity of the management requirements targeted by the TMN and OSI, the OSI management communication protocol is adopted by TMN as the protocol for network and network element management (ITU-T Recommendations X.710 [74], X.711 [75]).

OSI provides the DCN function by supporting a peer-to-peer communication paradigm facilitating Q3 communication. A set of seven layers full stack protocol suites for TMN Q3 interface are defined in ITU-T Recommendations Q.811 and Q.812.

OSI system management components CMIP (Common Management Information Protocol) and CMIS (Common Management Information Service) are the seventh layer communication protocol and service standards of the OSI reference model. CMIS and CMIP are the vehicles for conveying operations on objects, notifications from objects, results or errors of operations and notifications between management entities.

These standards support a set of common management operations which can be performed on managed objects or their attributes. These operations consist of:

- Operations on objects

  —`m-create` (confirmed)—The operation creates new instance of an object class with confirmation emitted as events.

—`m-delete` (confirmed)—The operation deletes object class instances with confirmation.

- Operations on attributes

  —`m-get` (confirmed)—The operation retrieves the values of attributes of managed objects with confirmation.

  —`m-set` (confirmed/unconfirmed)—The operation replaces the values of attributes of managed objects. Confirmation of the operation is optional.

  —`m-cancel-get` (confirmed)—The operation terminates linked `m-get` responses with confirmation.

- Other operations

  —`m-action` (confirmed/unconfirmed)—The operation invokes a user defined operation (methods) on one or more objects. Confirmation of the operation is optional.

Although an `m-get` operation can be cancelled, a general cancel operation is not defined. To cancel an operation which changes multiple objects status and attribute values requires extra technology support to ensure the atomicity of the operation. This poses two problems:

- This requires the transaction processing technology (or potentially the distributed transaction processing) to supply commit and rollback functions.
- It requires a semantic definition of the commit and rollback behavior on the managed objects. It in turn means that the definition of transaction semantics on the managed physical elements is required.

The technology and tools are not generally available to provide these supports. Therefore, the model does not allow the cancellation of other (writable) operations.

CMIS operations can be either confirmed or unconfirmed. For confirmed operations, responses or error reports from the agent are required to be sent to the manager who initiated the operations.

By defining the CMIS actions, the management applications can define their own operations which are outside the standard CMIS operations. However, the information agreement must exist between management entities before an action can be performed. This requires that every action be defined using GDMO and the action ID and information syntaxes to be registered with management entities on both sides.

OSI system management standards also support notifications to allow agents to report occurrences of predefined conditions in the form of `m-event-report`. An event report can be either confirmed or unconfirmed. The meaning of an event report is defined by the managed object notification specification. Similar to the actions, the types and the syntax of notifications need to be predefined using GDMO.

2.6.4 System Management Functions

System management functions (SMF) define the semantics and abstract syntax of the management information transferred between management entities for certain system management purpose. A set of system management functions has been defined by the ITU-T Recommendations X.700 series.

An example of such management functions is the event report management function. A managed object called *event forwarding discriminator* (*EFD*), can be created by a management entity. The EFD is used to determine which event reports are to be forwarded to a particular destination during specified time periods. A set of conditions can be specified as the forwarding selection criteria for the event reports generated. These conditions also specify whether confirmations are required.

The management platform is required to support the transformation of such management information. For example, during a connection setup, the management platform should support the negotiation between two management entities to decide which of the system management functions are required and supported.

2.7 BUILDING MANAGEMENT APPLICATIONS

Although Figure 2.5 presented a sample TMN–based system, the development of a real TMN–based management solution is a far more complex issue. OSI technology provides a system architecture for application development and deployment. It simplifies the development effort by allowing the common management functions, peer-to-peer communication functions, and object modeling tools and techniques to be separated from the application development. This separation of functionality has led to the commercial tools and platforms for OSI application development. The application development only needs to concentrate on the development of managers and agents and the integration of these management entities to form applications. In the next chapter, we focus on the development of the OSI managers and agents. We also discuss alternative architecture strategies to deploy these managers and agents in management applications.

3

OSI for TMN Network Management

TMN defined a general framework for managing telecommunication networks and services. Although this framework can be potentially supported by many different technologies, in the element and network management areas, TMN did adopt the OSI network and system management technology as the basis for implementing different components.

TMN management functions, especially element and network management functions, are essentially defined as sequences of cooperative activities between two management processes. One process is acting on behalf of the managing system, or applications, and the other is acting on behalf of the managed system, or network resources. This mode of operation fits very well with the OSI management systems.

In OSI network management (defined in the ITU-T Recommendation X.730-740 series), management activities are defined as a set of cooperative activities between a manager and an agent. The manager provides an interface to either management applications or network operators for management instructions. The agent has access to management information base and network resources for directly performing management functions on network resources. The communication between a manager and an agent uses the OSI systems management services and protocol. Such services and protocol are referred to as CMIS and CMIP (common management information service and common management information protocol). A CMIS and CMIP interface based on the full seven-layer OSI stack is used to support the Q3 interface in TMN.

Given the importance of the OSI technology, we will discuss in more detail the OSI-based network management for TMN solutions. We will discuss in this chapter the OSI services, protocol, object model definition method, interfaces, and main characteristics. More importantly, we will focus on how OSI technology has been used in the development of TMN network management applications.

3.1 CMIP AND OSI FOR NETWORK MANAGEMENT

OSI technology is jointly developed by ISO and ITU-T and adopted by TMN to provide the element and network management functions. The technology has many features which are designed to meet the unique requirements for managing telecommunication resources.

3.1.1 Communication Model

OSI network management technology supports a peer-to-peer, connection-oriented communication paradigm for managing telecommunication resources. In this paradigm, the resources are represented as managed objects. Two management entities are interconnected using this peer-to-peer communication paradigm to accomplish the management tasks. To communicate with each other, an initiating management entity needs to set up an association with the responding management entity first. This association is a logic connection between two entities and is supported by the Association Control Service Element (ACSE). The names and addresses of the management entities involved in the association need to be clearly specified. This is the reflection of the peer-to-peer nature of the communication paradigm.

An OSI management entity can have multiple associations with many of its peers. Multiple associations can also be set up between two management entities. An association identifier is used for each established association. This communication model is illustrated in Figure 3.1.

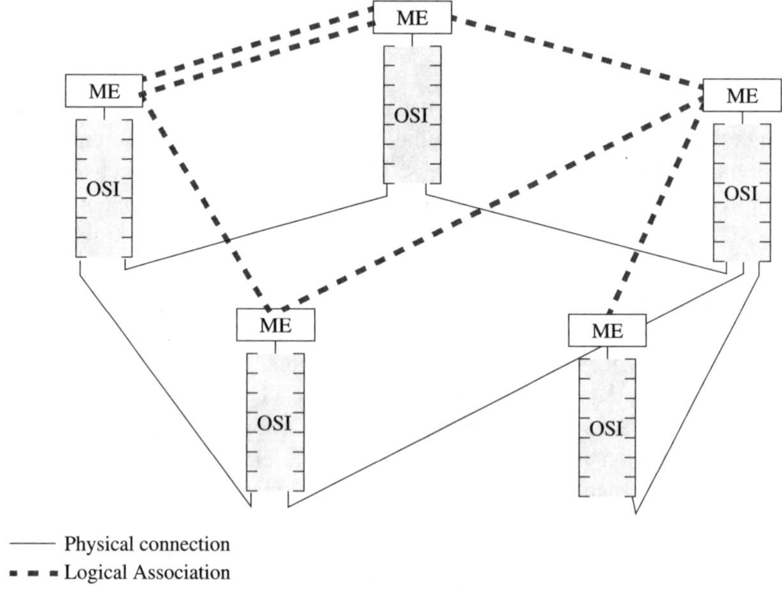

Figure 3.1 OSI communication model.

When establishing the association, a negotiation is carried out between the management entities to agree on the features of the association. The features to be negotiated include:

- The application context information corresponding to different *Application Service Element (ASE)* and access control related information.
- The service capabilities supported, such as whether to support the filter parameter, whether to support multiple object selection and multiple reply, and whether to support linked reply. These service capabilities are designated as functional units. During association setup, the functional units supported by both manager and agent are the functional units supported by this association.

- The types of *System Management Functions* (*SMF*)[1] supported. Each management entity involved in the negotiation needs to specify which system management functions are supported and for each function, whether it is going to play a manager role, an agent role, or both. The use of *System Management Functional Unit* (*SMFU*) negotiation allows a management entity to specify itself to be a manager-only, an agent-only, or a manager-and-agent in the applications.

An asynchronous communication protocol is used between a manager and an agent, that is, the manager sends a request to an agent, the agent performs the required operations, and then sends back a reply to the manager. The reply can either be a result or an error. A single request may be performed on multiple object instances from which multiple results may be generated. To support the asynchronous remote operations, OSI uses the *Remote Operation Service Element* (*ROSE*) as the base format for *Common Management Information Service Element* (*CMISE*) operations. Once an operation request is issued, the manager process is released, so it can perform other activities while the agent is processing the request and preparing for the responses. An operation identifier and linked identifier are used by both the manager and the agent processes to associate the responses with a request.

A management function is carried out between manager and agent processes. The manager process sends management requests (based on a set of management functions) to the agent, requesting certain operations on some managed objects. The agent informs the manager of any changes made to the managed objects which the manager is requesting. These requests and notifications (events) are formatted using CMIP protocol and transmitted by using the CMISE communication stack. This is illustrated in Figure 3.2.

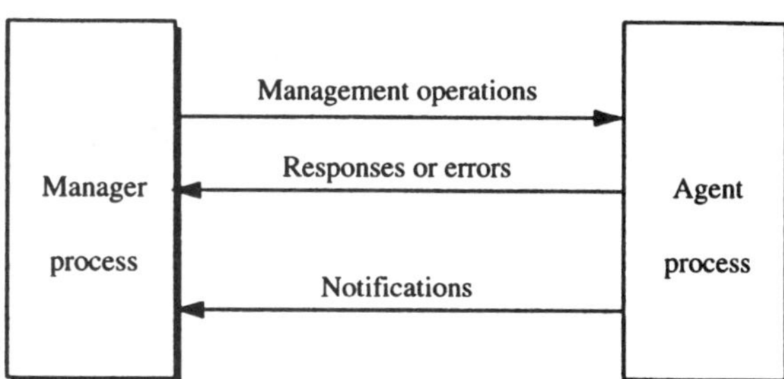

Figure 3.2 CMIS communication protocol.

3.1.2 Management Information

Managed objects in OSI network and systems management are defined in GDMO [79]. GDMO is an abstract language for modeling managed objects. It provides templates for the object class definition and uses ASN.1 for data type definitions. An OSI-managed object has the following features:

- A managed object is an abstraction of a physical or logical resource and is defined for the purpose of being managed by a management entity. An object has a set of attributes, may initiate notifications, and is the target of management operations.
- A managed object's behavior defines the effect and constraints of operations on objects and reflects the management view of the underlying resources.
- Managed objects that share the same attributes, operations, notifications, and behavior are grouped as an object class.

[1]System management functions are defined in ITU-T X.730+ and ISO/IEC 10064.

- An object class can be derived from another object class (superclass) and it can inherit the features of the superclass, such as attributes, operations, notifications, and behaviors.
- Object encapsulation is not enforced in the GDMO object model. All attributes of a managed object are visible to management entities and access to attributes can be made directly by managers.
- Some common methods which can be applied to any managed objects are defined as CMIS operations. Examples of CMIS operations are: create object instances, delete object instances, get attribute values, and set attribute values. In the modeling stage, a managed object needs to specify which of the object level operations, such as create and delete objects, are supported by the managed objects. In addition, the attribute level operations, such as get and set attribute values, are also specified at the modeling stage.
- Operations which are not common to all managed objects are defined in the form of *Actions*. An action is invoked through a CMIS common operation `m-action` with the action identifier as one of the parameters.
- Both common operations and actions are not invoked directly on a managed object. They are invoked through the management entity (agent) which is responsible for the target managed object. The target object identifier, the attribute names involved in the operation, and the action names are supplied as parameters to the common operation invocation.

GDMO also defines other facilities, such as attribute groups, packages, and name-binding which are required for object definition.

3.1.3 Management Information Base

Managed objects are controlled by an agent. Object instances are stored in a *Management Information Base* (*MIB*) which is a tree structure known as the object containment tree. The containment tree starts with a root node, and each object instance is a node in the containment tree.

The GDMO modeling language allows a name-binding definition for object classes. A name-binding definition specifies the valid location of an object instance in the MIB when it is created. With this facility, a dynamic MIB structure of object instances can be constructed.

Each object instance is identified by its object identifier which is a combination of the object class identifier and the object instance identifier. In a global management application, object instances are always identified by their *Distinguished Names* (*DN*s). The distinguished name of an object instance is the distinguished name of its parent object instance plus its own name which is referred to as the *Relatively Distinguished Name* (*RDN*). A distinguished name uniquely identifies an object instance and its location by traversing the relatively distinguished names sequence from the root. The object instance naming not only provides a unique name to an object instance, but also provides the navigation path to the object instance. This feature makes the object location nontransparent from management applications.

Figure 3.3 illustrates the structure of an example MIB and the managed object instances in the MIB. To construct the containment tree, the agent uses the GDMO name-binding rules to place the newly created object instance at the correct location in the containment tree. Once an object instance is created and placed in the MIB, it can be deleted but cannot be moved. Moving an object from one location to another can be simulated by deleting the original instance and creating a new one.

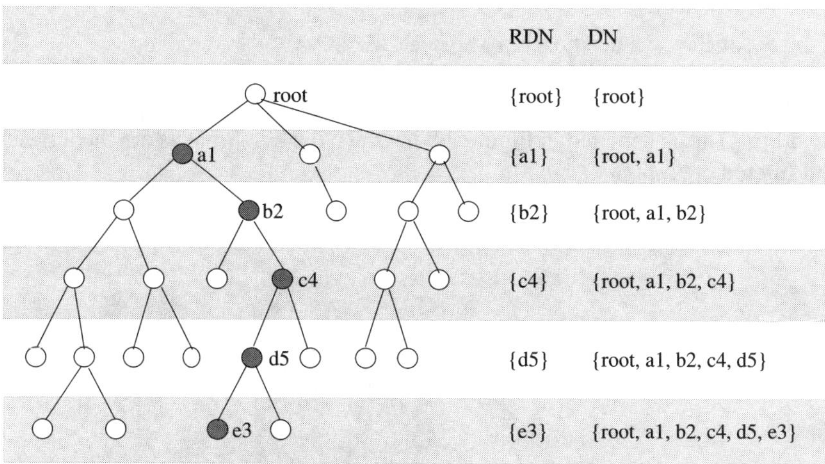

Figure 3.3 MIB structure.

3.1.4 Scope and Filter Operations

With the MIB organized in a tree structure, a single CMIS operation can be performed not only on a single object in the MIB, but it can also be performed on a group of object instances as a *bulk operation.* This is known as a *scoped* and *filtered* operation. A CMIS operation can specify a scope by defining a portion of the subtree in the MIB as the target object group. The subtree is identified using the base managed object as the root. A scope can contain the base object only, all objects at the *n*th ($n \geq 0$) level from the base object, all objects within the *n*th ($n \geq 0$) level from the base object, or the entire subtree from that base object. When *n* equals 0, a scope reduces to a single base object. Figure 3.4 illustrates the scope concept.

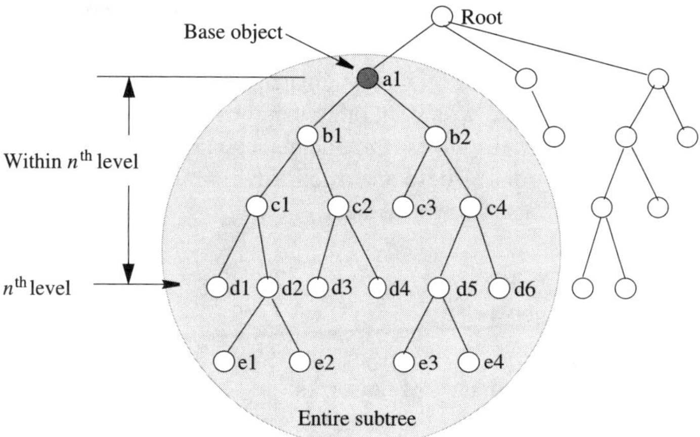

Figure 3.4 Scope supported by CMIS/CMIP.

A CMIS operation can also define a filter in conjunction with a scope to further identify a subset of the managed objects within the scope. The filter is a Boolean expression constructed from the following test conditions on attributes:

- Attribute value equality
- Substring value matching

- Presence
- Attribute value subset or superset operation

The CMIS operation is then applied to all objects within the scope and satisfying the filter condition. This is depicted in Figure 3.5 in which the dark objects are the targets of the scoped and filtered operation.

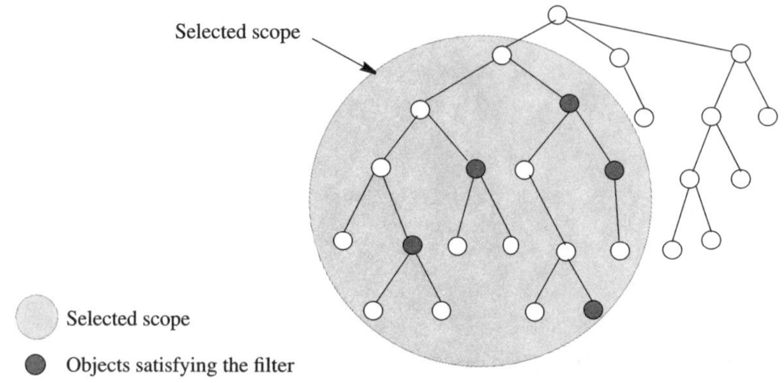

Figure 3.5 Scope and filter operation.

Because a scoped operation can be performed on multiple objects, these operations may produce multiple replies. In CMIS/CMIP, each of such replies is referred to as a linked reply. A linked reply is sent back to the manager as multiple responses.

At the time when a manager initiates an association with an agent, the manager and agent may negotiate whether scope, filtering, and multiple object selection are required by the manager and whether they are supported by the agent.

3.2 BUILDING NETWORK MANAGEMENT APPLICATIONS

The core activity of building TMN network management applications is to build management entities which communicate with each other using CMIS/CMIP in order to manage network resources. These network resources form the managed environments which are the targets of the management functions. These managed environments consist of a large variety of telecommunication infrastructures such as:

- Switching networks
- Transmission networks
- ISDN networks
- Broadband-ISDN and ATM networks
- Data networks
- IN networks
- Mobile networks

In telecommunication network management applications, there are two types of management entities, the *manager* and the *agent*. A manager is normally part of the TMN operation system which access agent functionality on behalf of the OSS and the applications. Agents typically reside on the network resource side providing the management functions on behalf of the resources.

Management entities (sometimes also referred to as management processes) form the main body of the management applications participating in global network and systems management activities. A management application has at least one manager process and one agent process. The manager process functions may be invoked by other telco OSS systems and may provide support for operators. The agent process is in control of the managed objects in the MIB and has access to physical and logical managed resources.

Many other management components contribute to the management applications; these include:

- OSI-based network management platform which supports CMIS and CMIP and uses OSI seven-layer reference model. This platform allows the manager to issue management requests to agents and receive the responses from the agents. It also allows the agents to issue notifications to the manager to inform any events or alarms generated by the managed network and network elements.
- Systems management functions which are the logical groups of management capabilities for a specific systems management purpose.
- Object-oriented management information model which is the logical description of the network and network resources to be managed. It provides the common understanding of the management functions and managed environment between managers and agents.

Figure 3.6 shows the general relationship between different components in a management system.

As indicated in Figure 3.6, the TMN network management system has a clear separation of the manager side and the agent side as the main management software components.

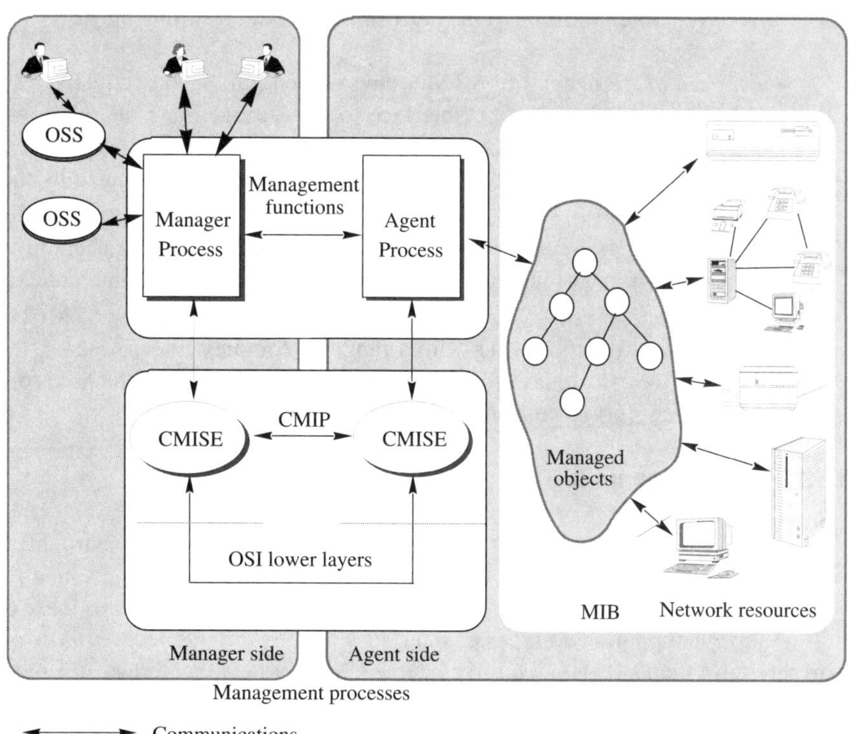

Figure 3.6 TMN management system.

The agent side contains all the network resources to be managed, which include physical network equipment, subnetworks and networks. It also contains the agent processes developed for these managed resources. The agent processes and the MIB they are managing are normally deployed close to the network resources. In many cases, agent processes are built on the managed network equipment as embedded systems.

The manager side often contains a management process which may be deployed as part of the applications and operations systems. The manager processes may also support interfaces which can be invoked by network operators.

The development of an OSI-based TMN management system is the development of both TMN manager processes and TMN agent processes. As we will explain further, the design of these management entities and the proper understanding and the use of manager and agent relationship will affect the architecture and the overall performance of the management systems.

3.3 OSI AGENT DEVELOPMENT

The development process of an OSI agent is the key to the success for building OSI-based management applications. The agent development is concerned with the capability of communicating with managers for management requests and performing these requests on the managed resources. In a TMN environment, an agent is required to support Q3 interface.

Currently, a number of agent development tools are available. These tools take the GDMO definitions as the input and automatically generate an agent template. The agent template mainly provides the CMIS/CMIP-based communication capabilities and some of the common containment tree management. Any system or device specific functions need to be developed as part of the agent development effort. With the help of agent development tools, OSI agent developers still need to address the following issues which can affect both the success of the development effort and the performance of the resulting agent:

- *The role of an agent*—In a TMN environment, an agent normally plays two roles. One role is to provide a Q3 interface to the manager, and the other is to act as a Q adapter (as defined in the TMN architecture) to provide an adapter function to non–TMN resources, so that these resource can also be managed by the TMN management system.
- *Support for CMIS operations*—This is the internal processing capability required for the Q3 interface, which is used to support agent–manager communications.
- *Object management*—Each agent manages one or a number of MIBs. This object management function deals with a range of complex issues such as: the MIB interpretation, concurrency control, containment tree management, scope and filter process, conformance, and device complexity management.

3.3.1 Role of Agents in TMN

As indicated before, an agent process in a TMN management environment has to provide two major functions. One is to provide a Q3 interface to managers, and the other is to provide adaption functions to non–TMN resources. Figure 3.7 illustrates these two roles.

An agent supports two types of interfaces. One is the interface to TMN operation systems through a manager. It is required that the CMIP-based Q3 interface be used for any communication with the manager.

The second interface an agent supports is the interface to the network resources including both logical resources and physical network devices. An agent is responsible for providing a unified view of the managed resources. It achieves this by encapsulating the propri-

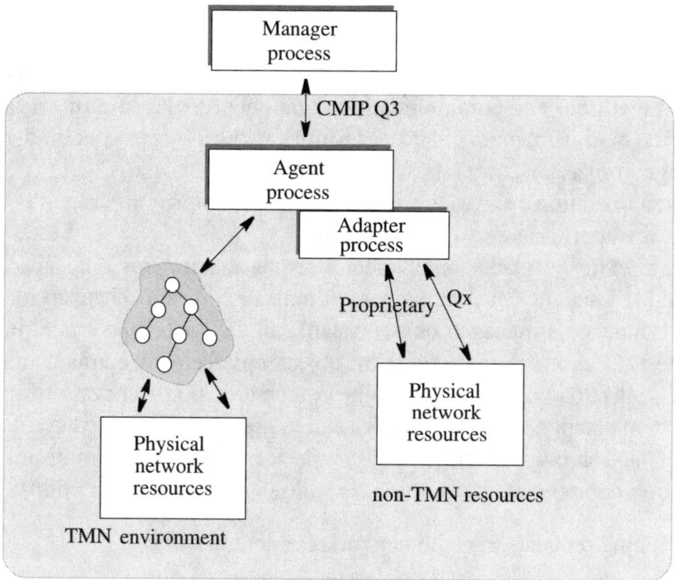

Figure 3.7 Functional role of OSI agents.

etary interfaces used by any network resources. The protocol used for the communication with network resources can be different depending on the interfaces supported by the resources. For a resource in a TMN environment, the interface can be one of the Q interfaces (e.g., Q3). For a non–TMN resource, the interface is likely to be a proprietary. For any resource not supporting Q3 interface, the agent has to providc an adaption function for it. In this case, a GDMO object definition for the managed object needs to be defined and mapped directly onto the proprietary interface.

Internally, an OSI agent may consist of three main components: communication, object representation, and adaption. This is depicted in Figure 3.8.

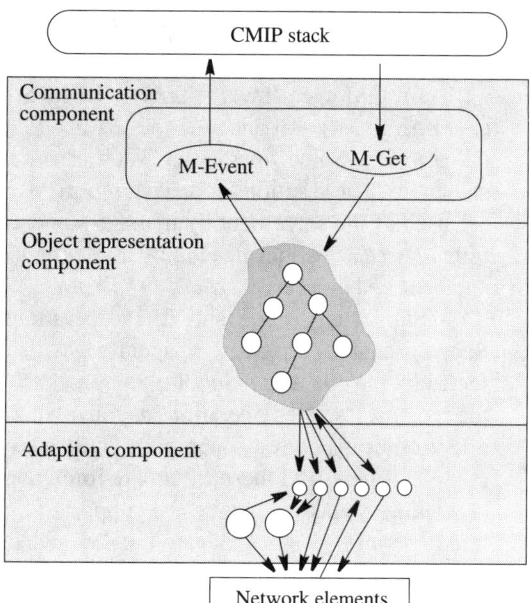

Figure 3.8 Systems component of an OSI agent.

The communication component is responsible for the initialization of the CMIP stack and the communication with managers. After receiving a CMIP request from a manager the communication component maps this request to an appropriate managed object instance in the containment tree. The communication component makes use of services provided by the object representation component to determine if the objects specified in the request (e.g., base-managed object instance) exist and whether the request is valid (i.e., attribute supports the requested operation). When the requested operation is complete, it is also responsible for generating any appropriate responses.

The object representation component is responsible for providing an accurate representation for both the MIB and the network element. It must provide a mechanism for accessing the attributes of each individual managed object instance in the containment tree. In order to handle the CMIP requests (such as get, set, action), the object representation component will use services provided by the adaption component. The object representation component uses the services of the communication component to send replies and to generate unsolicited event notifications.

The adaption component is responsible for providing a simple interface to the object representation component. It supports the following two main functions:

- mapping requests into the network element interface
- interpreting response data from the network element

Using an agent development tool, the effort required for the development of the communication component can be greatly reduced. This leaves the development focus more towards the implementation of the other two components: the object representation component and the adaption component. Typically, the adaption to specific network elements will normally require the majority of the work in building a Q3 agent, because this component handles the translation between CMIS object representation and proprietary object representation, and the mapping between CMIS operations and operations supported by the device.

Figure 3.8 only shows a CMIP get request and a CMIP event message generation. These demonstrate two important concepts: (1) a single request may result in multiple interface calls to the network element; and (2) the objects in the containment tree may send unsolicited event notifications to the manager (e.g., `stateChange` notifications).

3.3.2 Support for CMIS Operations

An OSI agent must support all CMIS operations (`m-get`, `m-set`, `m-action`, `m-create`, `m-delete`, `m-cancel-get`). It must also support CMIS notifications (`m-event`). Some operations (such as `m-action`) and notifications have very complex semantics. These actions and event reports need to be predefined in GDMO. These definitions are used as the agreement by managers and agents to understand the semantics of the operations. These semantics need to be implemented by an agent and supported by both MIB management and devices.

Supporting scope and filter operations is always an important issue for an agent. Scoping enables a request to specify what managed objects in the containment tree are affected by the operation. Scoping increases the complexity of the agent implementation due to the fact that a scoped operation may involve a large portion of the containment tree and it may affect other concurrent operations. OSI also allows CMIS operations to specify arbitrarily complex filters, and these filters are implemented by the managed objects.

Filters may or may not be supported by a device. For device supporting filters, the filtered operation can be delegated to the device. For those who do not have filtering capability, the agent has to be able to perform the filtering on behalf of the device and perform the operation on the filtered objects only.

Scoped and filtered operations provide OSI management entities with a capability to handle bulk operations efficiently. This capability can be very useful. For instance, a single `m-get` or `m-set` operation will allow an operator to retrieve or change the current status of all interfaces on a switch. However, when not used properly, it can be a source of inefficiency in the applications. For example, a user may unknowingly issue an operation through a manager with a very loosely specified scope which may involve many more objects than intended. The current CMIS standard supports only the cancellation of a get request using the `m-cancel-get` operation. To support this operation, the OSI agent must stop sending any further `m-get` results to the manager. However, cancellation of other requests are not supported by the CMIS/CMIP standards. This is mainly because of the synchronization problem—not many devices are capable of supporting the concept of rollback of a writable operation. Without this capability, it is impossible to treat scoped CMIS operations as atomic ones.

When handling an `m-create` request, the processing of superior object instance and reference object instance needs special attention. Instead of directly providing a distinguished name for the new object, the manager can specify a superior object as the parent of the new object instance in the MIB and a reference object instance as the type of the new object. For an agent, the issue is how to handle the case in which the object referenced does not exist, or the object instance is not of an appropriate type. In fact, the required object instance not existing in the MIB, or its type not matching the one in the request are the common problems faced by all CMIP operations especially `m-set`.

As part of the agent development, the ability to support multiple concurrent CMIS operations, made by different managers, is very useful. The ideal mechanism for supporting concurrency is the use of threads. However, the development cost for multiple threaded agents is higher than single-threaded ones. In particular, the ability to support multiple-write operations is crucial to increase the agent scalability and system throughput but adds more complexity to the development. The agent must have sufficient capability to maintain MIB data consistency and integrity.

3.3.3 Object Management

The object management of the OSI agent development addresses the following issues:

- *Device complexity*—which is concerned with whether the device (element) supports the facilities the agent requires to provide a suitable MIB interface and whether the device supports certain filtering.
- *Containment tree management*—which is concerned with what type of operations can be carried out on the containment tree and what properties must be maintained by the containment tree.
- *Object registration and naming*—which is concerned with the appropriate naming scheme for object instances maintained by the agent and the association of these names and the device.

All these issues impact on the ability of the development effort to implement the desired agent functionality. This section discusses some of these issues.

3.3.3.1 Device Complexity. The majority of the work involved in implementing an agent is the adaption component that talks directly to the network element. Often this work involves a protocol mapping from a standard interface to a proprietary interface. The interaction with the device and the availability of a suitable mapping is usually the key to the success of an agent development. To achieve these requires a detailed knowledge of the physical interfaces used by the device. Most device interfaces are complex and require special knowledge that is not common to all the agent developers.

For example, some device may have certain capability of applying some types of filters to the information represented using its own data model (e.g., to return all counters with values greater than a given value). Other devices may have no such capability. It will be very inefficient for the agent to implement filter operation without devices supporting filters. If the agent cannot delegate filters to the device, it can only achieve this effect at the MIB level which will place a heavy load at the device interfaces.

The data models used within the managed devices are not always consistent or clearly understandable. The granularity of data provided by devices is usually different from one to the other. Sometimes, it is hard to obtain specific data from a device. Instead, one can only obtain a large chunk of data, and tries to isolate the required specific data from the large chunk.

Device interfaces sometimes provide no guarantee on reliable and consistent services. In these cases, the agent has to provide some mechanism, such as caching or demand polling, to provide a reliable service.

3.3.3.2 Containment Tree Management. The management of the containment tree is one of the key issues the agent developers have to be concerned with.

First of all, for a management system, a persistence storage must be provided for the object containment tree, so every time when the system is restarted, all the management information can be restored. Every installation of the object containment tree should provide the persistent image of the containment tree, according to the distinguished name (DN) of each object instance. An important reason for keeping a persistent MIB image is that many managers may have already had their local view of the complete or partial MIB image, and all the management activities will be based on their local MIB view. Any change to the containment tree structure must trigger the notifications to all managers who have or used to have associations with the agent.

An agent must ensure at all times that only valid containment relationships exist within its containment tree. This implies that the agent needs to have the knowledge of the MIB containment relationships.

Containment management also includes performing normal management operations to all the managed objects. Examples of such management operations are:

- For each managed object in the containment tree, an agent must be able to find it according to its DN.
- Given a managed object instance, the agent must be able to locate its children nodes, and their relative distinguished names.
- When receiving a management request, such as `m-create`, `m-delete`, `m-get`, `m-set`, etc., the agent must be able to perform the operation on the given object instance in the containment tree.
- For a scoped operation, the agent must be able to identify all the objects in the scope and be able to perform filters on these objects.

One of the issues, as we have already indicated, is how to maintain the data consistency and data integrity of the object containment tree. Once an operation is required on multiple objects, the agent has to have the ability to coordinate the operations on these objects. Transaction Processing (TP) is one of the technologies that can be used for this purpose. Some other technologies, such as object locking, may also be used to support concurrent requests. This will support the isolation of object manipulations by the agents, such as the addition of nodes into the containment tree without them being visible to other managers before the operation is complete. The locking mechanism will also serialize multiple updates to the same managed object to safeguard the MIB integrity.

The management information maintained in the MIB must be synchronized with the information on the device. That is, the information in the MIB reflects the exact state of the network element. This is best implemented in a threaded environment where a background thread can be used to poll the network element, to ensure that the containment tree is an accurate reflection of the state of the network element.

It is also imperative that the agent maintain a valid representation of the device and also maintain a conceptual link between the managed object and the device concept it represents.

Figure 3.9 illustrates a containment tree management example. The figure shows two requests: a scoped CMIP `m-get` request and a CMIP `m-create` request. These two requests may be received from different managers at same time. In this example a managed object instance can be referenced in two different ways, and semantics of both requests are quite complex.

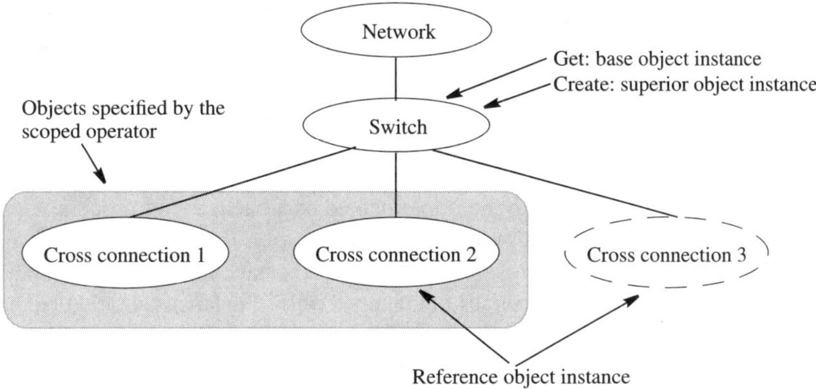

Figure 3.9 Containment tree management example.

The `Switch` object in the diagram is involved in both the `m-get` request and the `m-create` request. In each case its role is different. In the `m-get` request it is specified as the base object instance from which the scoping rule is to be applied (1st level subordinates are highlighted in the diagram as being the result of the scope). In the case of an `m-create` request the same `Switch` object is specified as the superior object instance. This means that the requested object to be created should have the `Switch` object as its superior object instance.

The `Cross Connection 2` managed object instance is also specified as the `Reference Object` instance in an `m-create` request. This means that if attributes in the newly created object do not have values specified in the request they are initialized with the values of the `Cross Connection 2` object.

The use of superior object instance and reference object instance are just two examples of variations possible when specifying CMIP requests. The semantics of CMIP requests can be quite complex as there is quite a large range of variations to a single CMIP operation and the effects these operations can have on the containment tree are wide ranging.

The number of managed object instances involved becomes an issue due to the functionality of the containment tree. All agent operations in an OSI environment are in effect operations on the containment tree. The efficiency of scoped and filtered requests is dominated by the efficiency of the containment tree implementation.

The containment tree in an OSI agent will usually be reasonably large. In all OSI agent implementations, there is a cutoff point where it becomes impractical to maintain the containment tree purely in memory and a dynamic loading scheme is required. Such a scheme prevents the binaries from getting too large and swapping.

The size of the containment tree also becomes an issue in a multithreaded agent as some operations, especially the scoped and filtered operations, may require an exclusive lock on a large number of managed object instances, resulting in delayed processing of other requests until the lock is released.

3.3.3.3 Object Instance Naming. OSI managers can and should have knowledge of all object instances which they have been informed by agents via object creation notifications. This information is crucial to enable the managers to map its presentation in user applications into objects in the agents MIB.

For OSI agents, it is important to ensure that the object instances in the containment tree are an accurate reflection of the devices being represented. In order to easily validate the object instances in the containment tree with the values contained with the device, the object instances must maintain some logical mapping into the relevant device concepts, for instance, a switch object has a conceptual mapping to its serial number in the switch.

The use of device-related names makes possible synchronization and validation of the containment tree. Synchronization involves the discovery of changes in the device state and the reflection of those changes into the containment tree. The synchronization and validation make the containment tree an accurate reflection of the device. Some agents will have different requirements as to how long it takes to achieve full synchronization with the device. Synchronization would usually be implemented as a background process within the agent. It is also possible that the agent may place higher priority on synchronizing some types of managed objects. For example, cross-connection objects have higher priority than log objects.

All internal object references to managed object instances should use the distinguished names instead of any programming variables or pointers. This enables the relationship to be restored after system shutdown and start-up. For example, this enables the containment tree to be reconstructed from persistent storage after the process failure.

Figure 3.10 shows an example of object instance naming. In this example, the managed object `ATMSwitch1` maintains two types of naming information. The first one is its distinguished name. To maintain the full distinguished name enables the object to be restored and it also enables a consistent image of the containment tree on both the manager side and the agent side.

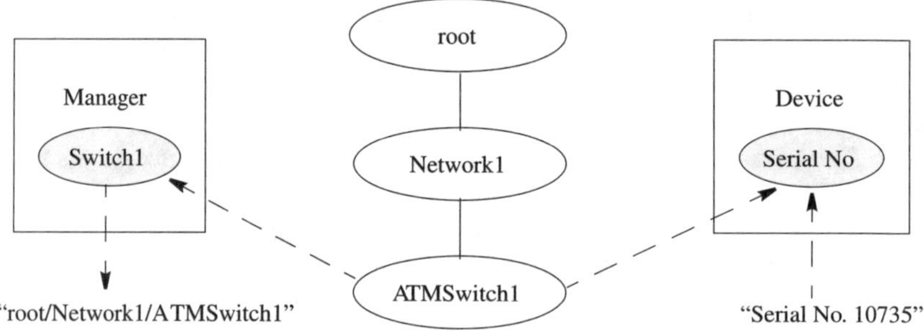

Figure 3.10 Object instance naming.

The other naming information is the relationship between the object instance and the concept it represents in the device. In the case of `ATMSwitch1`, it maintains a reference point to the serial number of the switch. This enables the agent to check the validity of the containment tree and possibly discover that the `ATMSwitch1` has been decommissioned and is no longer part of the network, thus the object instance `ATMSwitch1` must be removed from the containment tree. The removal of a managed object will trigger an object delete notification.

3.3.4 Software Reuse

Proper software reuse can reduce the overall agent development cost. Software reuse not only apply to the common components of an agent, such as the communication component. It also applies to platform related and device specific components, such as the object representation and adaption components. Often between multiple versions of the same type of network element there is significant reuse potential. In some cases, the differences between hardware models are only some extra performance counters.

In many cases the supporting routines for containment tree management and adaption can be reused in agent development to support a new network element from a different vendor.

3.4 OSI MANAGER DEVELOPMENT

Compared to the OSI agent development, the development of OSI managers are relatively simpler. The development of an OSI manager typically involves the following aspects:

- *Management interface*—As the OSI agent is responsible for performing the actual operations on managed objects, the OSI manager is responsible for providing access interface to network operators. The development of operator management interface is very similar to any interface application development.
- *Management applications*—Many management applications can be developed on top of OSI managers or sometimes developed as an OSI manager. The development of a management application needs to reflect detailed business requirements and objectives.
- *Management architecture*—This involves the strategic definition of the management application structure and relationships between components. This architecture should define the partition of roles and coordination between different managers, if the architecture identifies more than one manager.
- *Management protocol*—Similar to the agent development, the manager side also needs the support of the communication stack using CMIP protocol. Different from an agent, a manager initiates management operations and receives notifications.

In this section, we focus on the issues and experience of developing management system architecture, rather than focus on the development of individual management applications.

The OSI manager and agent concept is a modeling concept rather than a software architectural concept. This concept does not restrict a management entity to be a manager or an agent. Rather, it is a role-based concept which allows a management entity to play a manager role or an agent role for a particular invocation of a management function. In other words, it allows a management entity to act as a manager for certain management operations and act as an agent at the same time for other management operations. How to make use of the manager and agent concept may lead to different management software architecture options.

3.4.1 Centralized Management System

The conventional TMN development has the agent processes on the network resources side and manager processes on the operations systems side. It is therefore very easy to come up with a centralized management system with a manager communicating with a number of agents, as illustrated in Figure 3.11.

In some cases, an agent is developed and deployed for each managed resource, such as a switch and a router. In some other cases, an agent is developed and deployed for a group of network resources, such as a subnetwork or a network.

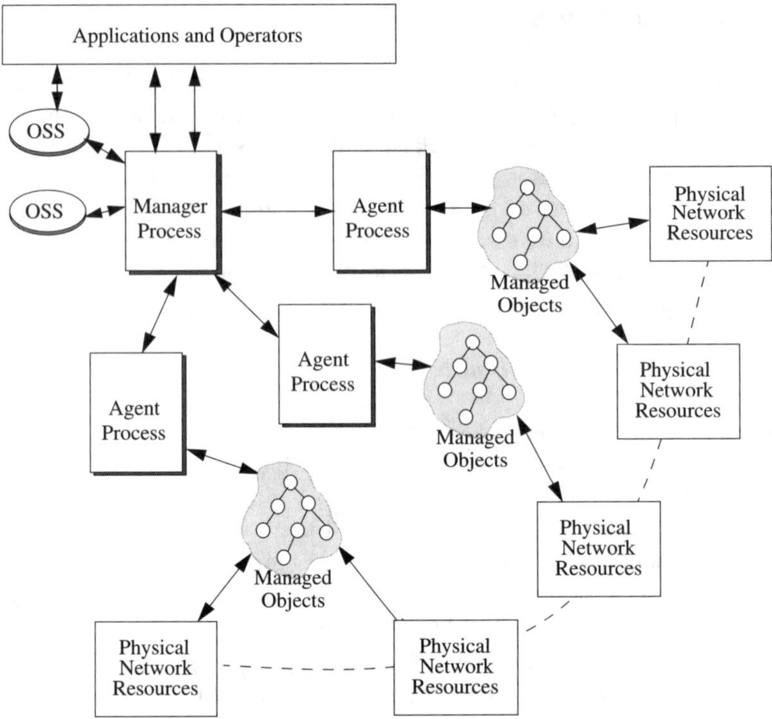

Figure 3.11 Centralized management system.

In a centralized management system, the manager provides the management interface to applications and operators and has no direct access to managed resources. It communicates with all agents for exchanging management information and relies on such information to decide the management operations that need to be performed by the agents.

The advantage of such a management system is its simplicity. Having access to the manager gains access to the management information of the entire network. This type of architecture is suitable for managing a small-scale and simple-structured network. The architecture can be easily supported by the traditional client–server software architecture and tools. For a large network, the manager might become a performance bottleneck, as it might not be able to cope with the communication demands with all the agents.

3.4.2 Manager of Managers Architecture

Different from centralized system, the management application employs a number of management entities to interact with the agents. These management entities can be classified as two groups. The first group consists of managers having direct communication with the agent processes. These are traditional managers. The second group consists of the management entities which do not have direct communication with agents. They only communicate with other managers (in both groups) and treat these managers as their agents. They are referred to as *manager of managers*. During management activities, a manager of managers sends a management request to other managers, and the latter send management requests to agents. The management processes in the first group act as agents when receiving requests from other managers, and act as managers when sending requests to agent processes.

This architecture supports multilevel abstraction. An example of the application is the layered management of a large network. A network to be managed can be modeled as having a network element layer, which contains all the physical network equipment, a number of logical subnetwork layers and a top network layer. Each layer can have a number of management processes developed for it. The management processes at the network element layer may be developed as agents. The management processes at other layers may be developed as managers. In this management structure, the subnetwork managers are managed by the network managers.

The layered manager of managers architecture is illustrated in Figure 3.12. This architecture is a reflection of the OSI peer-to-peer application paradigm. In a software engineering sense, each management entity is a component capable of being both a manager or agent. The role it plays is only relative to a specific CMIS operation and the role is only determined when the operation is invoked.

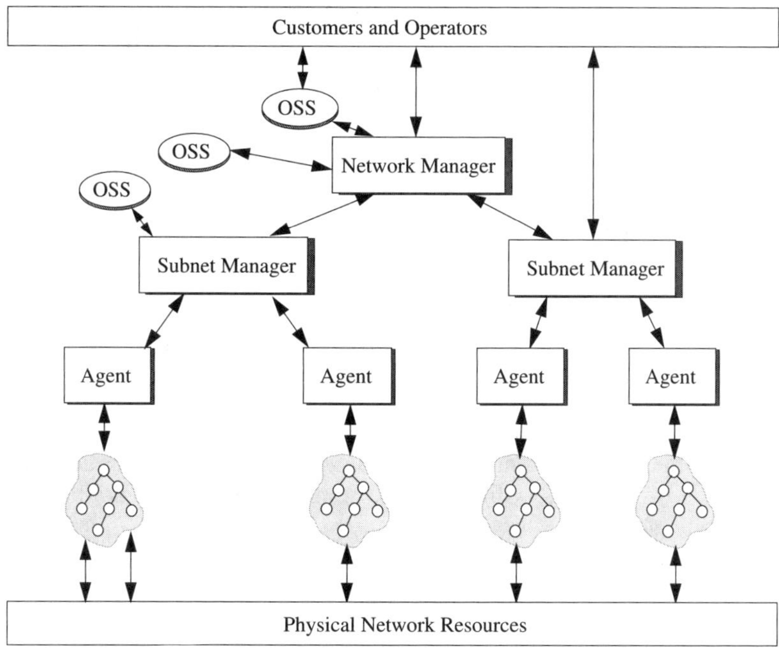

Figure 3.12 Manager of managers—layered management.

Another application of this architecture is to support functional separation in a management application that manages a large network. The management entities can be developed as specialized managers responsible for certain management functions, such as an alarm manager which manages all alarms generated by the entire network. A manager of managers can be developed to provide integrated management functions. This deployment is depicted in Figure 3.13.

It is also possible to develop a manager of managers system with a combination of layered management and functional management depending on the application requirement. A manager of managers architecture corresponds well to the multitiered client server software architecture and can be supported by those products and tools.

3.4.3 Decentralized Architecture

With the growth in size of the managed environment, the OSI based management applications move to more decentralized architecture and deployment. The requirements stem from the need to distribute MIB in multiple locations.

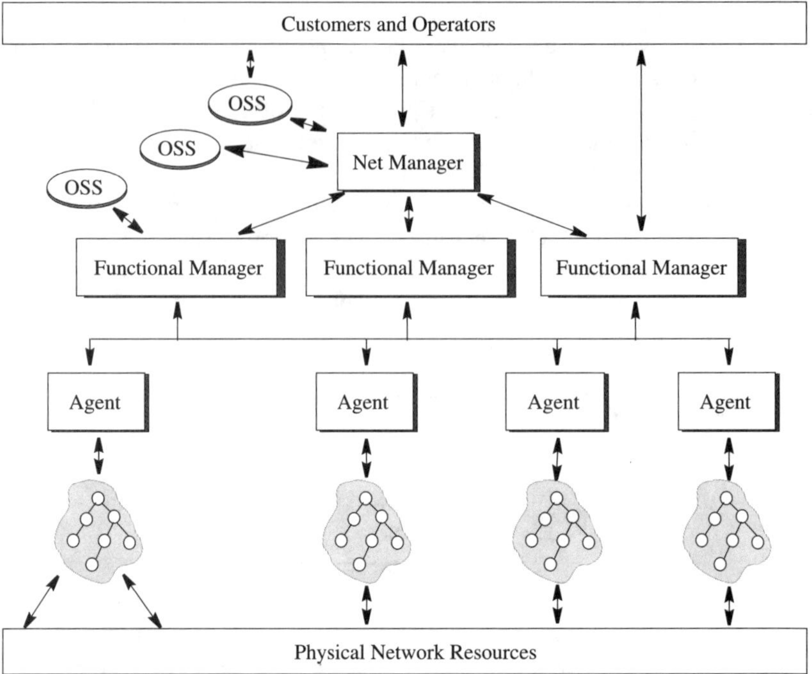

Figure 3.13 Manager of managers—functional management.

The main strategy of this architecture is to partition MIB into multiple subtrees and distribute them in a management network with different hosts. Each subtree represents a portion of the whole managed environment. This partition and distribution can be designed by using different deployment rules, with the most common one being to partition MIB according to the geographic needs.

Each subtree can be designed to represent a specific autonomous portion of the network infrastructure, such as a subnetwork, all resources in a regional operation center, or a large switch device with its dedicated agent software. Agents can then be developed using one of the two strategies:

- A specific agent for a predetermined portion of the MIB
- A general agent for the entire MIB but deployed at multiple locations to manage different type of deployment

The first strategy aims to producing a dedicated specific agent which may be more efficient. The strategy is more suitable for embedded systems managing specific devices. The second approach allows more flexible MIB partition and deployment.

With the MIB being partitioned and deployed on different hosts, multiple management entities can be developed and deployed according to the MIB deployment strategy. These management entities can be managers, agents, and manager of managers. They can be deployed in different hosts to collectively perform the management functions. The management application can be developed to coordinate management functions between all management entities. This decentralized system is depicted in Figure 3.14.

Each management entity in this architecture can be further decomposed into smaller units with different responsibilities and they can be deployed across the management network.

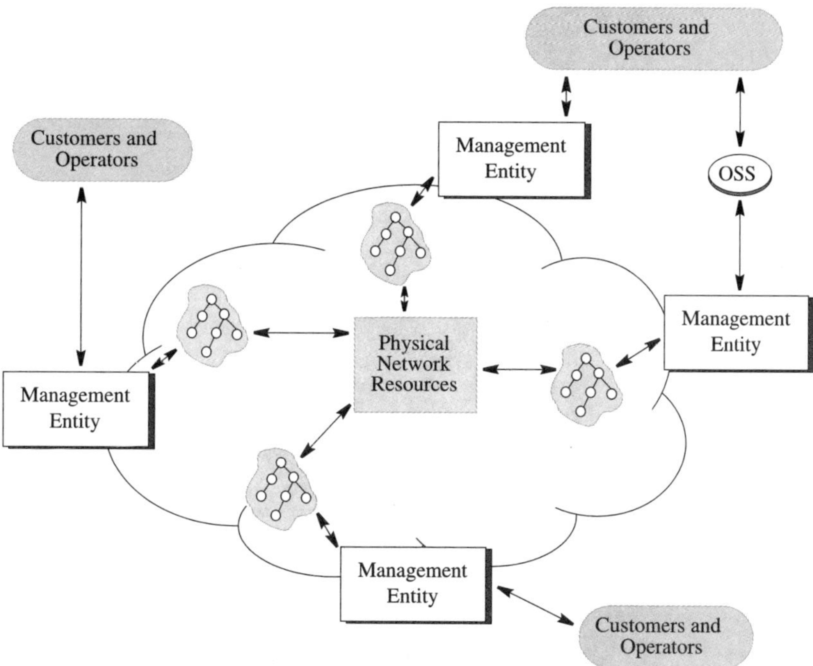

Figure 3.14 Decentralized management application.

3.5 BEYOND THE OSI NETWORK MANAGEMENT

Our experience of building OSI-based element and network management systems and applications has revealed some limitations of the technology. Today's business objectives to manage large global telecommunication networks require a more flexible and scalable solution framework to build and deploy management solutions. The process of management functional abstraction is essential to alleviate management applications from managing operational details of network resources. This allows us to build more scalable network management applications to manage larger networks and support more end-to-end network services. This requires the extension to the current technology framework which has been used mainly for implementing TMN element and network management functions. The mainstream IT advancements in areas such as distributed object-oriented computing paradigms and technologies have produced some clear technology candidates for these purposes.

The requirements for new IT technologies to be applied to telecommunication management framework are more obvious in the higher layer of TMN management. The software applications in service management and business management layers deal with more abstract concepts such as services, products, systems, customers, etc. They require more end-to-end network services and they manage more dynamic, large scale and heterogeneous environments. With the global deregulation and competition in the telecommunication industry, the industry-wide business process reengineering is shifting the business model from a resource-centric view to a service and customer centric view. This shift of business paradigm has posed strong challenges to the TMN development, and these challenges strongly influence the evolution of TMN technology framework and software engineering practices.

4

Telco Business Process Reengineering

The telecommunication industry, like all other industries, is driven by three external forces: customers, competition, and change. First, customers take charge in the customer–supplier relationship. That is, they decide what they want, when they want it, how they want it, and what they will pay. At the same time, their expectation of products and services soars. Second, competition intensifies. Competition now exists in different forms. The expansion of business into the information related market (e.g., data and information services) and the deregulation of the telecommunication market are the major causes. Finally, change becomes constant. With the rate of change of technology and regulations, change becomes unavoidable and it happens constantly.

The TINA-C has refined and extended these forces into six major trends specific to the telecommunication industry [93]:

- Diversification of market demand for services helps the individuals and organizations to improve their quality of life and run their business. The concept of service has been enlarged by the market demand to include more categories of services beyond the original voice and data communication.
- Services are available on a global scale and often integrating many administrative domains and technologies.
- Increased competition results in a need for cost reduction in all aspects of service and network operation and the ability to introduce and modify services quickly, cheaply, and reliably.
- Provisioning of services in a multivendor environment is required through interworking of equipment and software from different suppliers in a competitive environment.
- Changing regulatory environment has defined new responsibilities and conditions of interactions among competing service providers. This trend is moving toward a less regulated mode of operation.
- Telco customers demand increased management capabilities in the customer domain. This is caused by the increased degree of automation in the customer environ-

ment and the changing customer–supplier relationship. Telco customers demand more value from telcos to add to their own service offerings to their own customers.

TMF has also characterized the three driving forces as customers, service providers, and suppliers. The business challenge is to create the value chain between these entities, from the suppliers to providers and finally to customers. To do this requires the smooth flow of business processes and information between these entities. TMF sets this as one of the main objectives for the working program.

These trends drive the industry to change the way they provide and manage their services as the traditional ways are no longer adequate. A number of objectives are derived from these trends:

- Interoperability
- Reusability of application software
- Distributed execution of applications
- Support for new types of services
- Support for management
- Independence from computing environment and hardware
- Quality of service
- Scalability
- Security
- Compatibility with existing networks and services
- Flexibility against regulation
- Conformance

These objectives serve as guidelines for designing new services and their related computer applications.

4.1 THE REQUIREMENTS OF BPR

Today, telecommunication service providers face a number of major problems in providing and managing existing services, and in introducing new services. These problems are mainly due to:

- Legacy systems
- Data inconsistency
- Support for scalability
- Competition
- Organizational structure
- Inter-telco domains
- Management technology limitations

4.1.1 Legacy Systems

Telcos are dealing with a large number of legacy information systems on a daily basis. Many of these systems are built based on data models and technologies which are now out of date. If we don't have a sound strategy to the problem, with the rate of change of technology, a new system becomes a legacy system as soon as it is deployed. Legacy systems, therefore, cannot be avoided in a telco environment.

On the other hand, these information systems are operational systems (e.g., inventory, service design, billing, management, accounting) and they are mission critical. That is, the business cannot run without them. These systems also represent a significant investment by the telcos. As such, it is impractical and not desirable to completely replace them in one exercise of the business process reengineering. Instead, the requirement for the reengineering is to develop a technology migration framework so that these systems can coexist with the newly developed architectures and components.

4.1.2 Data Inconsistency

Legacy systems in the telecommunication industry were built to meet specific functional requirements and were implemented independent of each other. Many business services are encapsulated in these monolithic applications and implemented in heterogeneous forms (both hardware and software). As a result, data and fundamental functionality are unnecessarily duplicated.

For example, a telco has to interact with multiple OSS systems in order to provide a service for a customer. The OSS systems involved could be: inventory, service design, network configuration, and customer databases. At the same time, the process of provisioning a service typically involves manual operations in addition to updates of these information systems. In this process duplicated information about customers and services can be stored in these databases and applications. Additionally, due to various reasons (such as human errors and system crashes), this data can quickly become inconsistent.

4.1.3 Support for Scalability

The combination of rapid increase in size of management domains and the expansion of geographical boundaries of management activities leads to applications adopting a distributed architecture. The distributed architecture increases the scalability of these services and management applications. However, it also increases the complexity of these applications. Typical criticisms of distributed applications are:

- The applications are difficult to build and maintain.
- The consistency of data in a distributed environment is generally difficult to maintain.
- The applications are hard to manage.
- The applications are less secure.

In order to develop scalable applications to manage large and geographically dispersed systems, telcos need to address these problems by evolving the existing architecture using new approaches (of which new technology is a part).

In addition, the characteristics of these applications make it difficult to apply existing standard technology, e.g., distributed transaction processing (DTP). Some of the characteristics are:

- Nondistributed legacy systems involved
- Manual operations that cannot be easily automated
- Multiple transactional and/or nontransactional local systems involved
- Definitions of correctness and reliability (commitment semantic and recovery) have different interpretations from that of a traditional database-oriented transactional system

Also, in the area of customer service management, the Internet, Web, IVR (interactive voice response), and CTI (computer telephony integration) technology have increased customers' demand for electronic access to OSS systems for the purposes of service ordering, QoS management and billing information. The support for such electronic customer access functionality requires a high degree of scalability, security and in real time.

4.1.4 Competition

Another major problem is competition. Since the deregulation and expansion of business into data and information services, telcos face strong competition both from other telcos and from other information service providers. In order to maintain a competitive position in the market, they need to:

- Reduce time required to introduce new services to the market.
- Reduce costs in both existing services and introduction of new services.
- Enhance quality in terms of improved customer service and responsiveness.

All these requirements point to the need for a better defined and more automated business processes and the management architecture.

4.1.5 Organizational Structure

Historically, telcos are organized into a number of sections, departments and/or divisions based on the functions they perform or on the services they provide (e.g., accounting, broadband service provisioning, customer services, and network provisioning).

Applications and processes were designed by individual departments. In most cases, they were designed based on the functions they were supposed to perform with no vision on how they fit into the bigger process of providing services to customers. Thus, these applications and processes are internal function oriented rather than customer service oriented.

However, it is typical that the process of providing a customer service involves multiple management and information systems, and hence multiple departments. The process of providing a customer service now becomes a set of subprocesses in the individual departments involved. The coordination of the subprocesses among these departments is very difficult. In addition, each department have their own processes and information models, and thus, increasing the chance of data inconsistency occurring. This makes the customer service delivery and management inefficient and error prone. For instance, customer records and customer service profiles may be kept in different departments by using different databases. The business processes to access and maintain these databases are determined by the individual departments offering different services such as mobile, ISDN and Internet services.

Thus, to become a lean and smooth service provider with an organizational structure which is more customer service oriented, rather than internal functional oriented, is crucial to the success of the business process reengineering.

4.1.6 Inter-Telco Domains

As the services provided by telcos become more global, the size of the management space increases within and across telco domains. By and large, the coordination of processes across telco domains is much harder than that in an intra-telco environment. The difficulties are mainly due to the fact that different telcos have different hardware/software systems and they normally are governed by different business policies and government regulations.

Due to the deregulation of the telco industry, the wholesale interface becomes more important and more complicated. This interface is affected not only by the technology factors but also by the new business and legal framework in a global deregulated market.

4.1.7 Management Technology Limitations

In the technology arena, there exists a set of technologies, telecommunication standards, and industry standards on management functions of telecommunication networks. These include the ITU-T/OSI network management standards for computer networks and telecommunication networks, and the TMN standards specifically for the telecommunication network management. Historically, the development of these standards were strongly driven by network management focuses rather than the service management ones.

In these standards, the basic services and functions are well defined for building management applications. The communication between different components in a management system is also well defined. Examples of these communication protocols are CMIP and SNMP. There are a number of commercial network management platforms available for implementing management applications. However, these platforms do not address all the service management needs. In particular, they do not address the process and information flow from network management into the service management. This situation is mainly due to the fact that the requirements on these platforms are mainly driven by network management needs for historical reasons.

First, most network management platforms and toolkits mainly support management information exchange between managers and agents. Most of them support peer-to-peer communication paradigm, focused on the communication between a manager and an agent with clearly specified addresses, pre-agreed protocols, and well-defined actions. There is a lack of well-defined engineering solution architectures and commercial products that support distributed application development, and integration with network management environment.

Second, existing management platforms are based on network management standards and are developed for network management applications. As such, they do not meet the requirements of service management, nor that of business management requirements.

Finally, the customer service management, although not defined as a separate layer of TMN hierarchy, is becoming a crucial tool for telcos to stay ahead in the competition. A large number of commercial enterprise IT technologies are extremely useful to fill this space. However, the integration of these technologies with more traditional network management architectures and platforms still remain a difficult issue.

Based on these requirements, a new integrated management architecture and commercial products are required for developing service management and customer management applications. The platform should at least include the following features:

- Support both network management and service management functionality and the integration between two functions.
- Support integration with the customer service management and enterprise management.
- Provide better facilities for the development of distributed management applications.
- Enable management functions to be carried out across heterogeneous systems in a distributed environment for end-to-end management.
- Enable the rapid service creation and deployment.
- Support the technology migration road map.

4.2 THE MANAGEMENT OF SERVICES

Essentially, service management consists of management policies and mechanisms for services and relationships among services. It provides an integrated environment in which a service can be provided to customers, maintained, and reconfigured. The environment also allows existing services to be combined to form new services in a manageable way.

The principal roles of service management are:

- Customer care interface (the first point of contact with customers for all service transactions including provision and cessation of service, accounts, QoS, fault reporting, etc.) and interfacing with other administrations.
- Interaction with service providers.
- Interaction with network management layer.
- Maintaining statistical data (e.g., QoS).
- Interaction with the business management layer.
- Interaction between services.

Figure 4.1 summarizes the interactions between service management and its related entities.

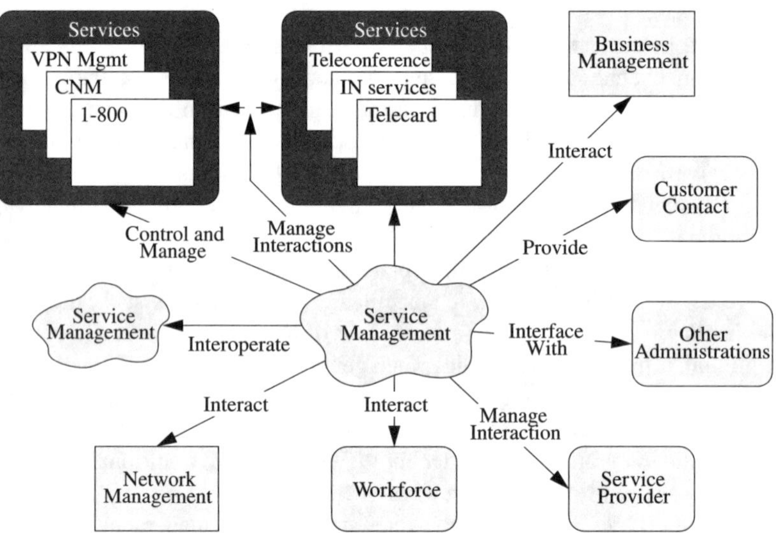

Figure 4.1 The TMN service management environment.

In reality, the existing service offerings from a typical telco include service management functions which are tightly coupled with the service offerings. The service management functions do not necessarily interact with functions in other services. Figure 4.2 depicts the current architecture.

Typically, every time when a telco introduces a new service, it builds a tailored network management system, charging system, and performance and fault management system for the service. These systems interact with other corporate information systems (e.g., customer billing system), but seldom make use of system components from other services. Not surprisingly, it is also impossible for other services to make use of these system components. As discussed earlier, the major factors that contribute to this phenomenon are: organizational structure, non-existence of integrated architecture, and technology limitations.

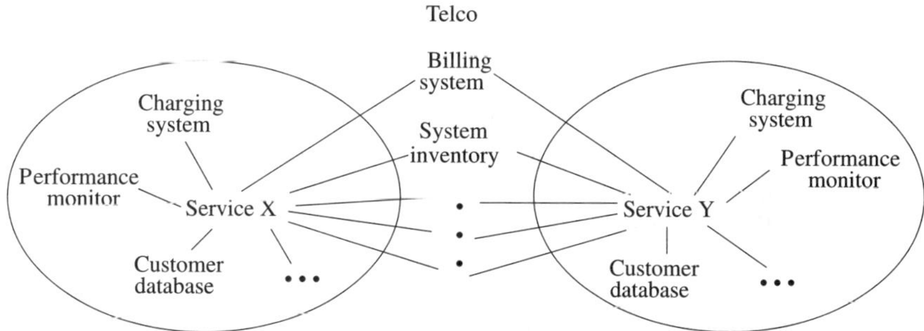

Figure 4.2 Current architecture of services within a telco.

Telcos are organized in sections in such a way that each section is responsible for providing and managing a service or a type of services. Conventionally, these sections are run independently. Each section builds all the specific management and operation systems (stovepipes) when a new service is introduced. These systems may have requirements to interact with systems in the corporate level, but not with similar systems in other services. Furthermore, technology limitations in the past make reuse of system components and distributed system architecture difficult.

The fact that a telecommunication network shares similar characteristics of a computer network provides the basis for highly automated service management. Technologies in managing computer networks can be employed to provide service management for a telecommunication network. However, as pointed out earlier, if we don't fully understand the business process flow from the customer requests to the service management and network management domains, an automated service management system is just impossible.

4.3 THE REENGINEERING OF SERVICES

The main purpose of business process reengineering is to move away from the network resource centric view of telco business to the customer centric view. The established infrastructure-based telco industry has already faced strong challenge from the new breed of service providers offering value-added services with better customer service management functions. It is important that the emphasis of the BPR is put on the increase of customer services with the support of end-to-end management.

4.3.1 The Broken Links

Derived from their definitions, the TMN management functions have quite different semantics in the different layers. The differences in semantics can be further illustrated by the following two examples.

While the network management technology detects and processes faults at the network level, there is generally a lack of mechanisms for these faults to be captured and interpreted at a higher level of abstraction—the service management level. The basic problem is that there is lack of real-time interface from network management to service management (and vice versa). Also service management applications tends to be non-real time and not designed to accept real time events.

Because of this deficiency, one of the problems being seen by the customers is that they do not get proactive fault (out-of-service) information on their subscribed services. The customers will not be told a service is unavailable (due to some unexpected faults at the network element level). Instead, they will find out only when they try to use the service.

Because service management technology is still taking shape as the next technology step beyond network management to meet the new business requirements, there is a general lack of well-understood definition of the interactions between service management systems and network management systems. Information about network elements finishes up staying in the network management applications and cannot find a way to propagate to the service management applications. For instance, a service management application cannot tell what services are affected when a network element sends a fault alarm.

As another example, the account management function in the network management layer is concerned with the usage of the transmission network. In the service management layer, however, the accounting management function is concerned with the usage of the service incorporating the factors such as tariff and regulatory policies. There is a lack of technology support to define these relationships such that the network usage information can be automatically translated into bills to the customers. Further, there is a need to support not only integration of management functions at different levels, but also to support integration of transactions. The accounting management function needs to support customer billing transactions involving multiple accounts and types of services.

Conforming to the TMN-layered management architecture, the service management interacts with the network elements via the network management layer in order to perform its functions. On the other hand, the network management layer propagates the network information to the upper layer when it is collected from the network elements. Therefore, the interaction between service management and the network management, as indicated by the shaded part 1 of Figure 4.3, needs to be clearly defined. This will allow generic software components to be implemented to automatically perform semantic mappings.

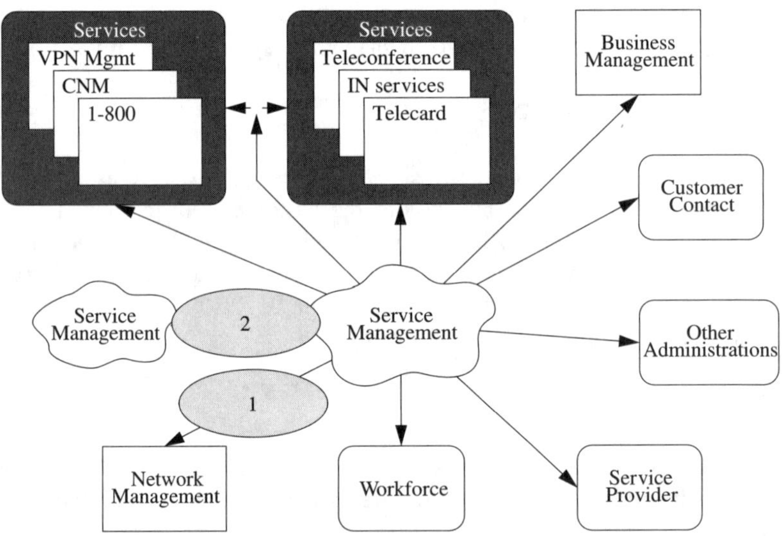

Figure 4.3 Missing links in TMN management environment.

Traditionally, service management systems are largely nonautomated. As a result, they cannot directly communicate with the service management systems in another domain (partners or competitors). The use of computerized information systems provides a better basis for them to directly communicate. However, the interoperability between service management systems is still a difficult task because of proprietary information models and different technologies that are used.

Even if they do communicate electronically in a common protocol, the communication usually takes place at the low level. For instance, two different switches may exchange billing data through a mediation device using OSI protocol. However, since there is lack of definitions on the contents of the information being exchanged at service management level, the automated business process and information flow, if it exists within a single service provider, stops at the boundary.

However, interoperability between service management domains (as indicated by the shaded part 2 of Figure 4.3) is such an important area for offering integrated customer services, or even to fulfill the legal obligations just to be able to operate as a service provider. When service management systems interoperate electronically and automatically with each other, a great deal of value-added services can be provided to customers. This interoperability also enables the management of a number of geographically dispersed TMN domains as if they were one. In addition, it enables the end-to-end services provisioning by a single service provider, particularly, a value-added retail service provider.

4.3.2 Support for TMN Service Management

Although the TMN model is not intended to be just focused on service management, the concepts and high-level objectives offer a general direction in solving the telco problems as stated in Section 4.1. The service management concepts provide a level of abstraction above the network management functions. The main objectives of the service management architecture are therefore:

- Define and view applications in the service level such that details of how the services are implemented can be hidden from the service applications.
- Provide customers with transparent services that go beyond the departmental boundaries and service provider boundaries.
- Reuse software components which provide the basic functionality.
- Maintain consistent data within telco and across telco management domains.
- Interact with network management services through a well-defined interface.

TINA-C, in addition to TMN, defined service management as a central component in the future (distributed) architecture of information systems in the telecommunication areas.

TeleManagement Forum defined a SmartTMN program to provide a more business-driven approach to implementing TMN. The center to this program is a detailed and concrete understanding of the telco business requirements and the mapping of these requirements to the business processes and technology independent agreements. The program provides not only a guide to the telco industry to conduct their business in a more competitive way but also a guide to procure technologies and develop solutions to achieve the business objectives.

4.3.3 Support for Network and Service Management Interaction

The key requirement for supporting interaction between service and network management is to provide a process and information flow-through between the two entities. This should include a detailed business process definition which maps the customer interactions with the service provider into the service and network management functions.

The map must reflect the natural flow-through, i.e., the service management functions must be well supported by the networks and elements. Further, the service management business processes must be technology independent. That is, it must support an end-to-end service view to the customers, and this service view must be cross network technology domains and underlying management technologies.

The key task to achieve these is an integrated process and information modeling from the business objectives (which is driven by the customer requirements) to the network resources. This modeling process must not be separated in layers and must be technology independent. The model must be supported by different technologies at different abstraction levels and integrated to support the whole business objectives.

The modeling process also provides us with a structured view of the business process integration. This view helps us to determine the integration strategy at the technology level. A good modeling process will not only define what must be done to support the integration, but will also lead to an implementation strategy focusing on what can be done now.

4.3.4 Support for Service Management Interoperability

The key requirement for supporting service management interoperability is to provide a transparent view to service management applications. This includes:

- *Object definitions and operations transparency*—it enables applications to access services offered by different service management domains in a uniform manner as they access the services offered by the local service management domain.
- *Coordination transparency*—it enables applications to use the same mechanism to coordinate service transactions involving both local and remote service components.
- *Communication transparency*—it enables applications to communicate with different service management domains using the same communication interface.

To support these transparencies, the service management interoperability can be achieved by providing:

- *Business process and information definitions*—This provides a common specification for the business processes and information models pertinent to the service management interactions between service providers. This common specification allows the applications in one domain to execute business transactions with the other service management domain in a predefined and automated manner.
- *Management of interconnection business transactions*—The mechanism used to coordinate local service transactions should be extended to incorporate interactions with peer service providers in the same transaction. This requires a transaction management component to be built as part of service management interoperability. This component should be provided by all service management domains such that the end-to-end service coordination can be achieved across the domains.
- *Support for business level communication*—The mechanism to support communication between different service management domains should be part of the service management interoperability. Service management domains should use well-defined business processes and information models to communicate.

4.4 THE FLOW OF PROCESSES AND INFORMATION

The critical task for the telco business process reengineering is for us to change the way we do business with our customers. The core of the telco service provider is services and the main objectives of a service provider are to provide services to its users in a fast, accurate, and low-cost fashion. These objectives are not isolated. They are embodied in the management of the life cycle of the customer interactions with the service provider which includes service enquiry and order creation and activation; and the smooth delivery of the services in a reliable, predictable, and controllable fashion. Furthermore, these objectives are not only supported by the customer care systems, they have to be supported by the service delivery and management functions existing in the OSS and the whole physical network infrastructure. Obviously, only a strong automation and smooth processes and information flow between multiple systems, in particular between service OSS systems and the networks can achieve these business objectives.

As discussed before, the current level of automation in the service provider's business environment is much lower than required in a competitive world. To search for a competitive edge and differentiation to deliver more value to customers, the service providers are actively engaged in the business process reengineering to increase the business process and information flow from the customer care systems to the network and elements, and flow among customer care and service management applications. These activities will result in an optimized operation environment in which service providers can run their business more efficiently and effectively.

The key success to this initiative is to model our business process accurately to correctly capture the requirements for integration and reengineering. This requires a hard and detailed look at our business to:

- Establish the global picture of the business processes. This enables us to understand the overall business objectives and our value proposition to our customers.
- Identify the functional components (subprocesses) which contribute to the overall business process. These components will be developed as building blocks in a more manageable fashion.
- Identify the value chain from the customer requirements and interactions to the network, and identify the network resources and information assets to meet these requirements and support these interactions. This process enables us to have a life cycle definition from customer care's viewpoint, rather than from isolated FCAPS functional viewpoint.
- Identify the information flow between processes. In particular, identify the integration points where the processes in different management domains are interacting with each other. Typically, the TMN network management functions are supporting TMN service management functions. This provides us with a key integration point.

This modeling process has to be conducted in a technology independent manner, as the resulting business process definition will have a much longer lifetime than the IT architecture and technologies employed to support it.

TeleManagement Forum has, through its wide collaborative work, developed a business reference model which is used as a reference point for the business process modeling. Figure 4.4 [99] depicts the business reference model.

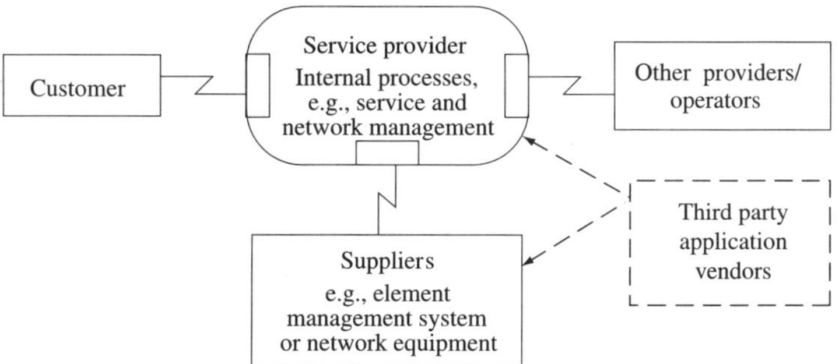

Figure 4.4 The telco business reference model (TM Forum).

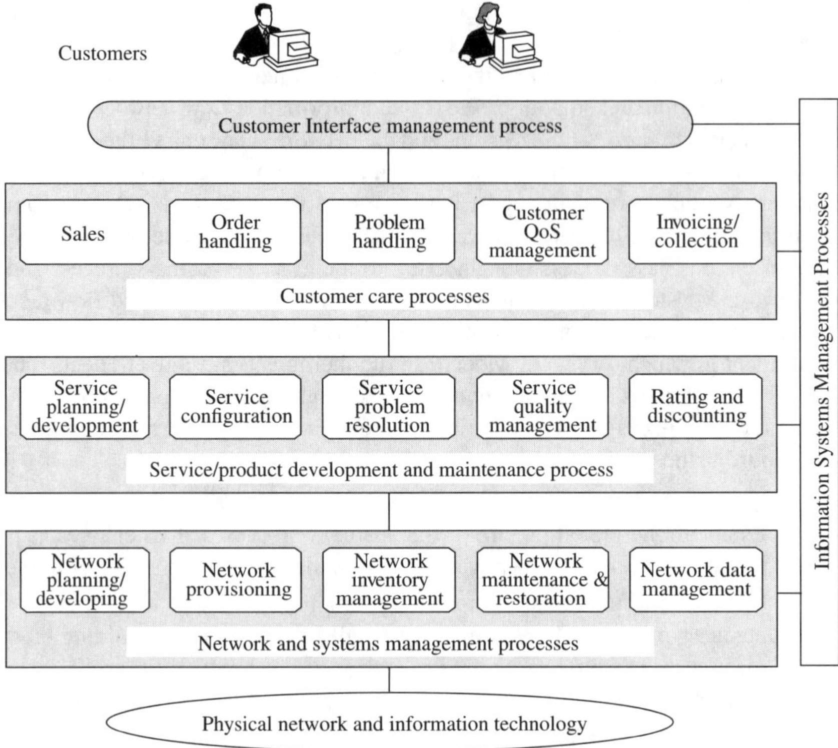

Figure 4.5 Business process model (TM Forum).

This simple reference model has clearly identified the management value chain, from customers to the service management and to the network management, including the externally sourced network assets and information and communication services. The reference model also reflects the business cooperations among service providers in order to fulfill customers' requirements.

Three interfaces are defined in this reference model:

- A customer care interface
- Supplier interface
- Inter-service provider interface

A customer care interface is crucial to the management of customer relationships. In a competitive environment, customers are constantly searching for the extra value added from their service providers. For example, customers require the kind of interactions which will allow them to integrate telco services to their internal OSS systems, such as integrating the telco billing with their own accounting system. They may also require the integration of telco services to the services they offer to their customers, such as E-Commerce.

A supplier interface provides the key to integrate service management and network management. This interface defines the mappings between the customer care and service management functions to the networks. Furthermore, this interface reflects the multiple technology integration requirements such that network resources must provide an end-to-end, cross technology domain view to the service and customer management.

The inter-service provider interface defines the framework of interoperation between different service providers in a competition and cooperation environment. It defines the behavior and extent of a service provider to interact with other service providers for business reasons (to support one-stop shopping for its customers, for instance) or legal reasons (to fulfill regulatory obligations such as FCC's OSS interconnection gateways).

From this reference model and the customer supplier value chain, a detailed business process model is developed by TMF to characterize the business activities and processes supporting the value chain. Figure 4.5 [97] depicts this detailed business process model as part of the Telecom Operations Map.

This model reflects a traditional TMN view of the telecommunication management. However, it separates the customer processes from the service management processes. This separation of processes is critical in business process reengineering. It puts a strong emphasis on the management of customers. It allows us to view the TMN model, in particular, the service management model, from a customer's perspective, rather than traditionally from a service providers' perspective.

4.5 THE SMOOTH OPERATORS—CUSTOMERS FIRST

In a global deregulated telecommunication environment, service providers are faced with unprecedented market competition to stay in the business. This has forced the service providers to reorganize business processes and adopt new technologies to manage networks, services, and their customers. Having better network management systems or operation support systems is very important, yet it is no longer adequate for service providers to stay ahead of their competitors. The service providers must seek differentiation in their management systems in order to win more customer business and loyalty.

The key functionality that offers this differentiation is the management of customer services and customer interactions. This management function focuses on the following customer interactions and activities:

- *Customer access to OSS systems*—This enables the customer to access their service records such as billing information, call records, QoS reports, etc. and current service subscription information in a quick and easy way, such as through Web and telephony.
- *Customer access to the new services*—This enables the SPs to roll out new services directly to their customers and allows the end customers to order new services, to have services provisioned, delivered, and paid on-line.
- *Efficient and prompt handling of customer requests and enquiry*—This allows the easy interactions between customers and SPs regarding to the service enquiry and requests.

In order to support these functions, the service providers must provide a more scalable, robust, secure, and flexible architecture, and to integrate this customer management and access framework into their overall management system.

Traditionally, the telecommunication industry follows the TMN model to partition the totality of the management systems and functionality. This has resulted in multiple management layers. The TMN model has further partitioned the management functions (activities) into five areas—fault management, configuration management, accounting management, performance management, and security management.

Whereas this may have correctly captured the features of different management activities, this model strongly represents a management view from an SP viewpoint. It reflects the internal view of a telco's network and systems, and it does not necessarily reflect how an end customer, whether a retail or wholesale customer, views the system.

The TMF's telecom operations map partitioned the telco's network and operation systems differently. It breaks down the management processes to:

- Customer service/access management
- Service/product development and maintenance
- Network and systems management

Also, from a customer viewpoint, any interactions between an end customer and an SP can be split into the following areas:

- *Service fulfillment*—it deals with the interaction with customer's enquiry, order, provisioning, and tracking.
- *Service assurance*—it deals with the interaction with customer quality of service, performance, discovery, reporting, and rectification of service degradation, submitting and tracking trouble ticket.
- *Billing for the service*—it deals with the usage data collection, billing record generation and management, tariff management and bill production, presentation, and payment.

This is known as the FAB model (fulfillment, assurance, and billing) as depicted in Figure 4.6 [99].

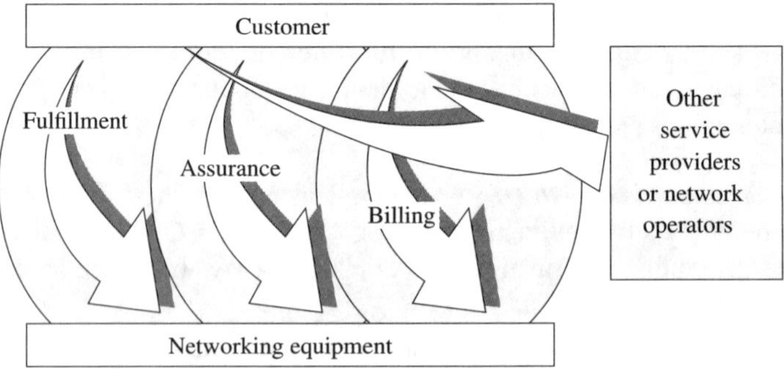

Figure 4.6 FAB—the end-to-end processes view (TM Forum).

Although this partitioning of management domains and activities is based on the underlying TMN model, it does reflects strongly the customer viewpoint and it focuses strongly on the customer management and smooth process and information flow from customers to the network and operation systems. This change of perspective in service management is vitally important to survive in a more and more deregulated and competitive world. It should be regarded as the key to the business process reengineering.

4.6 THE IMPACT OF IT TECHNOLOGY ADVANCEMENT

As discussed before the business process reengineering requires:

- A clear rethinking of the global deregulation, competition, and the business differentiation
- A clear understanding and definition of the business process flow-through from customer access to services and to the network resources

- A clearly specified information model which represents the business intention, describes the nature of the process flow-through, and integrates different components and functionality of the business processes
- An integrated IT technology strategy to support this business process and information model

The advancement in the IT industry in the past decade has provided a solid basis to support this BPR initiative. The maturity of the IT technologies and improvement in software engineering practice have made it possible to address these BPR requirements. It allows us to reengineer the technology solutions to support the reengineering of business.

From the perspective of supporting BPR, the key development of the IT technology includes:

- *The maturity of distributed processing paradigm, architecture, technology, and products*—One of the main technical requirements for BPR is to enhance the scalability (both in size and geography) of the OSS systems at a low cost. For instance, the billing enquiry system may need to support millions of customers nationwide to access their billing records with thousands of concurrent access requests. It is very difficult to support these requirements satisfactorily without the maturity of the distributed technology.
- *The maturity of the object-oriented technology*—After twenty years of continued research and development, the object-oriented technology has reached a significant level of maturity. The OO technology is not limited to provide a powerful tool to support business process modeling and information modeling, its marriage with the distributed technology provides a much powerful tool to specify, implement, and integrate large, distributed, and complex operation systems.
- *The technology to support application development*—We have also seen a huge advancement in the application development support area. We've been given more tools to develop system components and applications, more tools to integrate different system components to form solutions, and more tools to integrate the legacy environments. Above all, the emergence of the component-based development architecture will assist us to develop technology independent solutions to support business processes.
- *The user access technology*—The global phenomenon of Internet and Web technology has resulted in a much enriched environment for users to access OSS functionality. This has extended the means users and telcos can interact through Web, telephony, information kiosks, and many other means. This technology improvement has made it possible for telcos to provide more customized services to end clients. This also provides the basis for supporting the notion of customer self-service and micro-marketing by using a range of on-line access technologies to telco's service environment.
- *The maturity of the software engineering practice*—The theory and practice of the software engineering have matured considerably in the past decade. These have provided an important means to support the software development in the BPR era. The features such as:
  —distributed development
  —quality management and assurance
  —new process management paradigm to encourage the continued development and enhancement

  all contribute to the solution development life-cycle management of the telco OSS systems.

4.7 TECHNICAL CHALLENGES

The advancement in the IT has provided the strong support for the business process reengineering. This has, however, presented the software practitioners in the telco industry with new challenges. Providing business solutions to the BPR requirements is not simply writing programs, we have to deal with more complex issues:

- Understanding more complex business requirements
- The multiple technologies for supporting same functionality
- The integration of multiple technologies
- The different level of maturity of the technologies
- The complex technical solutions to support business requirements (such as distributed technology for supporting scalability)
- The complexity arising from the new solution architecture (security, management, maintenance, etc.)
- The more complex software engineering practice such as: distributed project management, component-based paradigm, life-cycle management, interface management, etc.

All these issues directly influence the result and the quality of the business process reengineering. They form part of the essential training process for our modern software practitioners in the BPR age.

Perhaps the most significant influence of IT technology to the telco industry's BPR initiative is the combination of distributed computing technology and the object-oriented paradigm. In a strict technology sense, most BPR initiatives in the telco industry involve building a large scale distributed system to support:

- ever-growing global communication network and new diversified network technologies and the demand for cross-domain transparent management
- ever-growing demand for new distributed services with more and more end-to-end capability and transparency
- increased customer access demand and the integration of telco services with the customer business environment
- desire to reduce the cost of business process reengineering by migrating to more flexible, adaptable and configurable management architecture with a better legacy integration and migration strategy.

This presents a challenge to the IT practitioners in the telco industry, an industry was so far largely based on centralized, non-object-oriented solutions, to migrate to the new technology environment. For many this migration process is not a happy one, this is the price of the business process reengineering.

5

Migrating to Distributed Systems

Picture yourself in this situation: you have been responsible for the roll out of the network and service management systems in the past. You have built some centralized systems which worked fine until one day your manager asks whether your system can support the network ten times larger than it currently supports; whether it can support twice as many applications and still run twenty times faster.

You quickly established that it is futile to argue with your boss that this is impossible, as he or she is in turn pressed by the upper management in this trend of business process reengineering to get every system (including people) to work more, faster, and cheaper. You asked around, checked the latest technology development, went to trade shows and forums and concluded that it is probably time to do something about your old systems. External consultants were called in to help; your colleagues and management were lobbied to fund your new initiative; internal R&D projects were underway to prove the new concepts, you were convinced that the migration to the new technology is essential. In today's technology climate, it is no surprise that you targeted the distributed object-oriented technology.

So you are ready to build your large distributed OO systems.

5.1 WHY BUILD DISTRIBUTED SYSTEMS

Distributed systems are hard to build. If the classic programming is one dimensional, the distributed systems are three dimensional. We use to deal with a traditional linear flow of application semantics. This is now complicated by multiple time stamps (nonlinear) in different parts of the system and the multiple spaces (naming, searching). The components of the system are starting and stopping everywhere; it can easily get out of control.

Distributed systems are very hard to manage. Similar to the global telecommunication networks, there isn't an overall management system that allows us to effectively manage the distributed systems. As we now have different components and subsystems running at different locations. The multiple locations introduce many complications such as naming, searching, and communication between components and subsystems. We have to manage different users of the system who may have totally different roles and may access the system for different purposes. We also have to manage different functions of the subsystems, such as some components act as confederates in a federated service deployment and some others act as backups or standby modules for a reliable deployment of some services.

The distributed systems are prone to more types of software and hardware failures. In addition to the potential software and hardware failure in a centralized system, we are now faced with the communication failure, synchronization failure, wrong timing sequence, violation of data integrity in different parts of the system, inaccuracy and inconsistency of multiple copies of the same data, security breaches, etc.

Many of the technology and process weaknesses and lack of maturity are really weaknesses in the state-of-the-art software engineering disciplines. They are inherited in application development, database design, network interface planning, etc. These weaknesses are not only related to distributed systems. However, the complexity in the design, development, and deployment of distributed systems makes it harder to overcome these technology and process weaknesses.

Then why build distributed systems? The answer is that most of us simply don't have a choice. If our environment requires:

- the system to scale to manage very large environment
- the system to have acceptable reliability at low cost
- the system to achieve a better component reuse
- the system to achieve extensibility of functionality

then we should consider distributed systems.

5.2 A SIMPLE MIGRATION

People responsible for building today's management systems typically have to go through a migration path. They were probably responsible for the existing, nondistributed, nevertheless still working systems. There is nothing wrong with the existing systems except their business managers keep asking the system to manage larger and larger networks, more customer records and try to add more PC applications, until the system cannot take any more.

A typical telco operation systems scenario is like this. We are likely to have a management system which resides on a server supporting most functions such as communication (proprietary, or SNMP/CMIP), issuing management functions to the network, running all functions to support applications and maintaining all application data in a database. It typically looks like the architecture depicted in Figure 5.1.

Now with the additional networks and applications we have to support today, we soon run out of system capacity. Our management system is slow and simply cannot cope with the

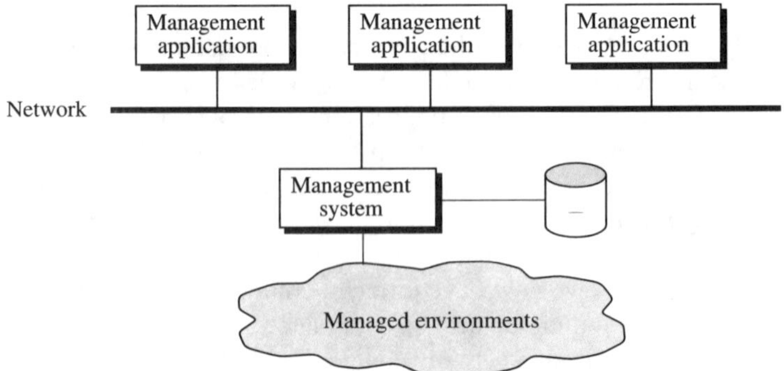

Figure 5.1 A typical management system architecture.

extra management functions required. We may have also experienced some problems caused by the original design deficiencies, such as the reliability is low. We cannot afford a Cray computer to run the software. Our first reaction is likely to add a second machine, and the new system may look like that shown in Figure 5.2.

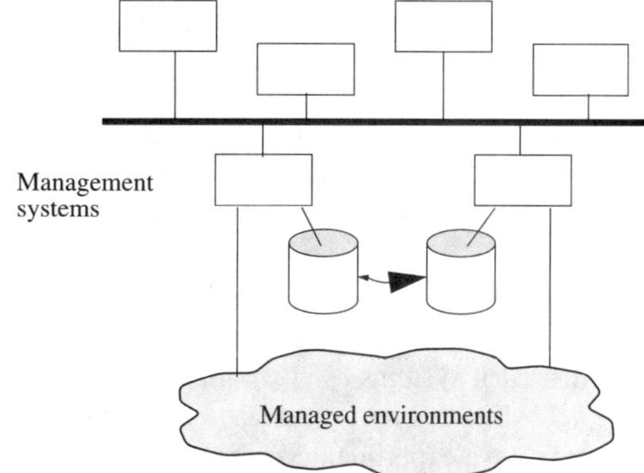

Figure 5.2 An extended management system architecture.

We now have a second server which helps to balance the ever-increasing system load, to replicate data to strengthen the reliability and to support more applications.

This trend will continue as our needs increase. Before long we will have to add more machines to the system, and it goes on.

Once we add an extra machine to the system, we start to experience the change of behavior in the software development. We now have two servers, for which we have to build some software to identify them. We now have to know where to store data, on which of the two servers? All our applications will have to know which server to connect in order to get specific functions. We have to design extra software to make sure that the two servers access the exact same state of information. It doesn't take us too long to realize that we are gradually building functionality that already exists in some distributed middleware, such as CORBA. After some investigations, experiments, and perhaps some external support, we finally decide to re-architect the operation systems to make use of the off-the-shelf products (or the product of another department in the organization) so that the architecture looks like the one depicted in Figure 5.3.

We are excited by the newly discovered vision and the initial success and we try very hard to convince the management to support us to build new distributed business operation systems using distributed middleware.

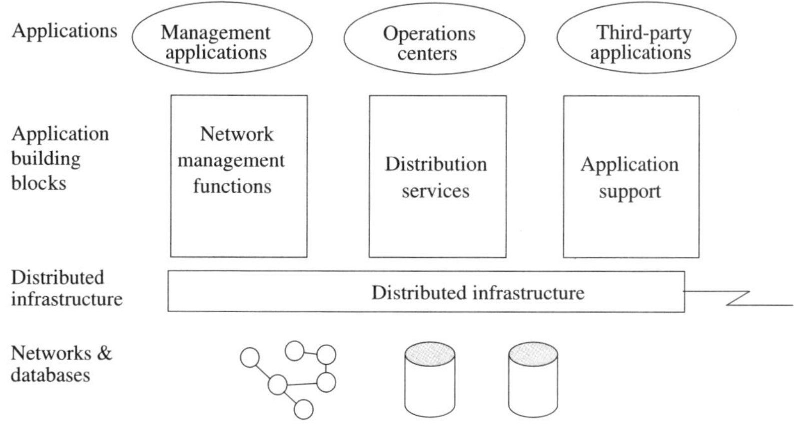

Figure 5.3 A distributed management system architecture.

Many of us have gone through this scenario of the technology migration. We don't have to be a genius to get this far. However, as there are many technical conservatives who refuse to take any initiative to embrace new technology advancement, there are also many of us who failed to realize the complexity and the cost associated with such initiative.

Building a distributed system is not hard using today's technology; but building a large distributed system which is scalable, reliable, and likely to work is hard. As the large distributed systems are very complex and most of us don't have the luxury to turn this exercise into a hugely expensive R&D effort, we need to know exactly what we will encounter when we set out to build large distributed systems.

5.3 COMPLEXITY OF LARGE DISTRIBUTED SYSTEMS

Now we are interested in building large distributed systems and we would like to know the complexity of building such a system, in particular, we would like to have some confidence that such systems can be built before we embark on the project.

The reality in today's telco industry is that some software engineers and practitioners may have experimented with the distributed systems, but they have not realized the complexity associated with building large and commercial distributed systems. As a consequence to this, many efforts turned into costly R&D projects.

What are the issues relating to the complexity of the system? While many issues are contributing to the complexity of the distributed systems, we list the following major ones to illustrate our view:

- The size of the system
- The scalability of the system
- The reliability of the system
- The extensibility of the system
- The maturity of the distributed technology
- Safety of the management solutions

5.3.1 Size

How large is a large system? How many objects are there in the system? The word object here is used loosely to mean the individually identifiable items in the managed network environment which are the target of the management functions. The examples of the objects include managed objects in the management information base which models the network and elements, the functional objects representing the system functions, and the entities representing the presentation aspect of the system such as maps, icons, etc.

Table 5.1 (Rose, [58]) presents an example classification of the size of the system. In this table, the size mainly reflects the order of magnitude of complexity.

From the maturity of today's technology and software engineers' experience of building distributed systems, the complexity of the systems can be qualified as follows:

- *Small systems*—These sized systems can be built successfully with today's technology and experience with very little learning effort.
- *Medium systems*—These sized systems can be built with a high success rate with today's technology and experience. Although some advanced system modeling and large distributed OO system experience is required.
- *Large systems*—These sized systems can be built, but little experiences exist. Most efforts reported today are of R&D projects. People setting out to develop such systems should allow a significant research component and resources.

TABLE 5.1 The Size of the Management System

Size	Example System	Number of Objects	Users	Operations per second	Machines
Small	Entry niche telco	0.5M	10	100	5
Medium	Cellular, ATM	5M	50	1,000	50
Large	Small PSTN	50M	500	10,000	500
Huge	Large PSTN	500M	5,000	100,000	5,000

- *Huge systems*—These sized systems require a major R&D effort in conjunction with building large prototype. Short-term success of building such systems is unlikely.

The main reason for this classification is:

- The current distributed OO technology is starting to become mature but not quite ready for real industrial strength solution.
- The distributed OO technology is very flexible and its architecture and design styles represent a new paradigm for people who are accustomed to architect and design traditional computer systems. The lack of experience in designing a working system is a real issue for the industry. This is the main reason why we only see many research and development type of effort being made by the industry, but very few real production efforts.

5.3.2 Scalability

Scalability deals with the issue regarding whether by adding more computing resource, the system can support proportionally more users and manage larger environment with the same functions and performance.

Systems can scale in two different ways. We can always buy a larger machine to support more users. If the computer resource/performance ratio is good, then this may not be a bad option. Given that most of us cannot justify a Cray computer, the options for us is really to employ a distributed architecture.

Before we embark on a project to build a distributed scalable system, we need first to understand the implications to the operation and management of such a system. At least the following three factors will be affected by these implications: the deployment, the users (developers), and the management.

First, the distributed system is not a shrink-wrap product. Its deployment could be a very complex issue. We cannot simply click the install button and expect the whole system to be deployed smoothly. A distributed system can be configured in totally different ways; it is very probable that some configurations do not perform well.

Second, the users of the system, in particular the application developers, have to be educated to understand how distributed systems are designed to work. The understanding of the large search and naming space, the distance and ways of configuration between components, and the deployment of different functionality of the system is very important to users and application developers.

Third, we will have to face the rise of the management cost for such systems. We will have to manage security since the whole system is divided into components and they are deployed in different physical locations; we will have to manage work groups, domains, large numbers of users, and the policies of their access to the system; and we have to manage larger naming space. More importantly, the distributed scalable systems are meant to grow bigger and bigger with increased managed environment and users, we have to manage the growth and expansion of the system.

Contrary to someone's belief or the advocacy of some of the zealous technologists, the distributed technology does not provide scalability by default. That is, we will not get a scalable system by simply using a distributed technology such as CORBA. Scalable architecture is a result of good design, not a product of a technology or a tool. However, the distributed technology is designed to provide functionality and services which can be used to support scalability. If these functions are used correctly, we can build scalable architectures and production systems.

5.3.3 Reliability

Reliability is the most vulnerable area of distributed systems. It is quite obvious that when the system gets distributed, the system becomes more vulnerable and it is far more difficult to enhance the reliability of a distributed system than a centralized system. In most cases, the reliability cannot be achieved by just adding reliable hardware due to the probability of the communication failure and the vulnerability of the distributed software.

It is important first to establish reliability classification and reliability needs for the system. There is no point to say that we want the most reliable system only to find that we cannot pay for it. Table 5.2 (Rose, [58]) lists four classifications of reliability relevant to the distributed systems by the effect of component failure.

TABLE 5.2 Fault Susceptibility Models (1)

Model	Effect of Component Failure
Vulnerable	May cause total system failure
Graceful	Failure limited to operations affected by the failed component
Tolerant	Operations in progress can fail, but can be retried successfully
Immune	No effect

As defined in the Table 5.2, the level of reliability a system exhibits is defined by the effect of a component failure to the whole system. A system is *vulnerable* if any component failure can render the whole system unworkable. A system is *graceful* if the effect of a failed component is limited such that only affected operations fail. An analogy in a standard operating environment of a desktop computer might be that the system is vulnerable if any operation failure will lead to the reboot of the operating system, whereas the system is graceful if by killing that failed process, the rest of the operating system can continue functioning properly.

A *tolerant* system still exhibits failure, but the failed operations can be retried by the system successfully. Very little user intervention is required except the operations to repair the crucial hardware failure.

An *immune* system does not exhibit any failure at user level. Normal operation will continue when a component of the system fails.

It is important to understand that the more reliability is required, the more complicated software design techniques and hardware strengthening techniques have to be used, thus the costlier the system. The following Table 5.3 (Rose, [58]) illustrates the software design techniques have to be used to achieve each level of reliability and the associated development cost.

Table 5.3 shows that for any level of reliability beyond a graceful system, we have to be prepared to pay a much higher cost for it. Therefore, we need to be very clear about what our reliability needs are, what system behaviors we are prepared to accept when a component fails, and how transparent we want to the recovery mechanism to be.

TABLE 5.3 Fault Susceptibility Models (2)

Model	Design Techniques	Extra Cost
Vulnerable	Do nothing	Zero
Graceful	Replication of meta-data and configuration data Unified service, partial execution	Minimal (20%)
Tolerant	Replication of instance data Replicated collections	Need to add significant development cost (100%)
Immune	Requires research	May quadruple development and ownership cost

For example, the distributed transaction processing technology (DTP) is generally used to enhance the data quality of a distributed system and to protect the data integrity in the events of system or component failure. The standard DTP services and software architecture provide complete and automated failure recovery (except in the certain events of secondary storage failure). It is an important technology to build a tolerant system.

The transaction processing technology deals with the following scenario. Part of the application semantics may involve a defined sequence of manipulation of some data resources (such as database). These data resources are held at different locations and are managed by different software components which collectively achieve the application semantics. The application semantics require a high degree of data integrity among the data resources and a high degree of reliability over component failure. This requirement is manifested by the following four points:

- Every operation within a defined sequence has to be successfully carried out; if this cannot be achieved, then none of the operations may be performed (such as debit and credit operations in a funds transfer scenario). This is known as *atomicity.*
- Every operation in the sequence, if performed, has to be performed correctly. This is known as *consistency.*
- During the sequence of operations, no other application may access the state of the data resources currently being manipulated. This is known as *isolation.*
- In the event of component failure within the sequence of operations, the system has to be recovered to the state before the sequence starts. This is known as *durability.*

The technology roughly works as follows in a simplistic case. The application first sends instructions to all components responsible for data resources to indicate the start of sequence. It then sends individual instructions to each data resource to perform certain data manipulation (credit, debit, etc.). It then enquires each one as to whether the operation is successful. If all answers are positive, the application asks everyone to make their data manipulation operations permanent and indicate the termination of the sequence. If any one answer is negative, the application asks everyone to abolish the update and restore the data to its original state and indicate the termination of the sequence.

Every participant in this distributed semantics has to log sufficient information for the purpose of recovery. When a failed component is resumed, it knows how the whole sequence is completed and performs its required operation accordingly.

However, when we use distributed transaction processing technology to build a tolerant solution, there are at least four factors which will influence our decision considerably.

- *Complexity*—The distributed transaction processing is generally regarded as a very complex technology. It adds considerable complexity to the system and requires more development expertise in the project.
- *Performance overhead*—Considerably more communication between components of the system is required to achieve the coordination, thus potentially slow down the system.

- *Application*—The full extent of the technology may not be applicable in certain domains. For example, how can we reverse the operation to its *original state,* when the laser beam is already turned on and data has already transmitted? How can we *lock* two switches while performing a path provisioning given that an operator sitting in front of a switch may take the termination point we just reserved?
- *Maturity*—The commercial technology might be mature but its integration with other technology (such as distributed transaction processing in a distributed OO technology) or its application in specific domains (transaction processing technology for network management [80]) may not be mature.

To analyze the needs is to make trade-off decisions. Such as, if transaction processing technology may not deliver required throughput, then can we use *store and forward* techniques (store all operations, release the application to perform further actions, and forward the operations to data resources) to increase the throughput, knowing that this may result in the small windows of data inconsistency. Do we need total transparent error recovery mechanism given its performance overhead and development cost? Can we live with relying on certain manual interventions, so long as the errors are logged, reported, and thus controlled?

This example illustrates that we must understand exactly our requirements and accepted trade-offs when we design reliable systems. This gives us the best utilization of our development resource and optimal system function to suit our needs.

5.3.4 Extensibility

A key benefit of migrating to distributed architecture is the enhanced extensibility. The pain associated with the legacy systems is always manifested through the lack of extensibility. These system functions are *fixed* as the systems are typically built as monolithic structures and any extensions to the functionality will result in rewriting of a complete new system.

Extensibility increases the complexity of the system, but its benefit far exceeds the cost. Extensibility is directly contributing to the business solution life-cycle management by supporting an easy and flexible function roll-out of the system. To design an extensible system forces the designer to think the issues beyond this current release and the current functionality of the system. It requires careful planning and designing in order to get the long-term benefits as the reward. It forces one to consider the following issues:

- Interoperability
- Flexibility
- Functionality roll-out
- Reuse
- Protect the investment by encapsulation
- Constructing building blocks rather than building systems

Interoperability deals with the relationship between internal components of the system and the relationship with the external systems. Interoperability should always be encouraged as support component-based development. This allows us to replace (or upgrade) certain functionality of the system components with a new one for the reason such as better performance or lower cost, or simply for the reason that the system can interact with more external systems to achieve the *plug and play.* Interoperability is an important aspect of extensibility. However, interoperability should be designed to have a right level of granularity. We should avoid overloading the system by introducing too many interoperability points, as plug and play often comes with an integration cost.

Flexibility deals with the level of effort required to build a component-based system and the effort to integrate these components. When we design an extensible system, we should always ask: Is our system allowing us to replace the database with another (better performed, cheaper) one? Can we change the ORB because the commercial deal with the current supplier has gone bad? Is the interface used between components easy to extend if required? Obviously, the answers to these questions depend largely on how we treat the issues of reuse, encapsulation, interface modeling, building blocks, etc. when we design the system.

We cannot build the complete functionality in one development effort. Thus we should always define carefully a functionality rollout of the system. We can have different (or a mix of) functional rollout strategies. We can build progressively more functionality into the system. For instance, an integrated network management environment may start with only a connection management function and it can progressively roll out performance and fault management functions. We can also support complete functionality with a weaker solution. For instance, an integrated network management function may start without an industry strength database support. No matter what approach we use, careful planning and overall design is crucial to the successful rollout of functionality. A disastrous scenario is that when the performance management system is to be provided in the integrated management solution, it is realized that the information model used by connection management to represent network connections cannot be extended to support performance management. Similarly, another scenario is that when an object-oriented database is to be added to strengthen the application integration solution, it is realized that the data access interface in every application is tied to the relational tables in the previous system.

Software reuse is one of those amazing concepts that everyone accepts as valid but which very few people practice or are able to practice. We are forced to consider reuse when we build an extensible system. The key to reuse is the appropriate specifications (interface and behavior) of the components. The difficulty associated with reuse is always caused by:

- We don't know what the component is doing—lack of specification, or
- We know what the component is doing, but we cannot use its interface, it is either too specific or is too general—specification is inappropriate.

Lack of proper specifications and documentation processes is the enemy of reuse. These systems really deserve the name of *legacy* system. They do something, but no one is sure of what exactly. These systems are always designed by people whose only goal is to solve *the* problem—connecting two ports in a vendor switch. Opposite to this are the people who always think that they are solving a problem which is a specialization of a larger problem—connecting two end points in a subnetwork. Smart designers will consider that someday someone will have a general solution for configuring any subnetwork connections and that our solution is only a special implementation of it. So they provide a full specification detailing exactly what they have (and have not) done. This solution may be more appropriately called a *heritage* system—they had their time, but still have their elegance and we can still benefit from it. When time comes for us to replace it with a WDM configuration system, this can be done without effecting much of the system interface above it. We have successfully protected our investment by focusing on interface and encapsulating the implementation.

5.3.5 Maturity of Distributed Technology

The past few years have seen the adoption of the distributed object technology in the telecommunication industry. A large number of R&D exercises have been reported. These exercises focused on the functionality of distributed object technology and were very useful to ascertain the value of the distributed object technology in the TMN solutions domain. In the service management area, the technology enabled us to build scalable distributed applications to manage customers, services, and computer and telecommunications systems. It also provided the component-based development environment to allow rapid functional rollout which gave telco a potential competitive edge in the deregulated market.

On the other hand, most reported efforts suggest that the industry is only experimenting with distributed object technology. There is lack of evidence that the industry has totally embraced the new technology in the operation supporting systems. One of the major concerns is the maturity of the new management technology.

It is evident that, while some of these technologies are rich in desirable features to implement TMN functions, they still lack the required industrial strength in many areas to support reliability, fault tolerance, high availability, real time, high degree of data integrity (through integrating with transaction processing technology), ease of use, and indeed the general level of product quality.

This observation results from our own experience. We were once involved in an R&D effort to investigate the CORBA-based distributed network management architecture, where the management functions such as connection management requires a high degree of data integrity. Obviously, we would consider the distributed transaction processing technology being used to provide the ACID[1] transaction properties. The research focused on the OMG Transaction service and its implementations found in the commercial ORB products. The result of the product evaluation was very disappointing. The CORBA products in the transaction area lack the maturity and sophistication found in the traditional On-Line Transaction Processing (OLTP) products by a large margin. We were also surprised at the general lack of the experience in the telco industry to use the object DTP technology in network management, whereas we all know that the DTP technology has been used to support relational database and client service computing architecture for a long time.

However, from the business management viewpoint, the industry is forever faced with this problem of dealing with immature technology. Everyone's management system is a mixture of out-of-date components (which are largely working but really need replacement); old components (which are supporting major operation functions, work reliably, fine-tuned, and stable, but largely boring) and new experiment systems (exciting, unstable, too early to support the whole operation system, but have great potential for future migration). All of us are constantly looking for the new evolution of our operation systems to the next new technology. We just need to learn how to cope with the management of technology evolution and the risk of it.

The distributed object oriented computing paradigm provides a good framework for managing this process. The paradigm encourages the software reuse and component-based development, which helps to manage the process of evolving immature technologies.

5.3.6 Safety of the Management Solutions

Safety provision of the system supports the recovery from failure, the enhanced availability of the system, and the consistent status of the system. The reliability of the system can be enhanced by the redundancy built in the distributed systems and the effective system management functions to manage the backup. It may also require a version of transaction processing or locking mechanism to safeguard the consistent status of data.

Due to the distributed nature of management solutions, it is required that the overall architecture should support the following features:

[1]Refer to the property of Atomicity, Consistency, Isolation, and Durability.

- The consistency of data, operations, and quality of services
- The maximum availability of services
- The continuous operation of component systems over network failure. In the event of the failure of some portion of the network, for example, the whole system is not available but the remaining networks and subnetworks should still be functioning.

5.3.6.1 Consistency of Management Data. In many existing management systems, there exist no provision of consistency of management data. This is manifested because:

- there is no guarantee of the integrity of backup data—we do not know if the backup data are completely in sync with the master copy
- there is no guarantee that the change over from master system to a backup system will not result in lost data
- there is no support of multiple writes due to lack of control for potential conflicts
- there is no mechanism for keeping data stored in database consistent with the data in the network elements

To develop a safe management system, it is required that a consistent data within management system must be maintained. In particular, the following features must be supported:

- the integrity of backup data must be guaranteed
- the change over from master to backup system must achieve maximum consistency—any data loss must be kept to a minimum and traceable by means of notification or log
- the management system must have sufficient support for multiple writing
- the database must be consistent with the network element states

5.3.6.2 Safety of Atomic Operations. Many existing systems provide no support for atomic operations. An atomic operation in service management systems shares some characteristics of an IT system, an operation requires updates on several databases in an atomic fashion, i.e., all or nothing. However, to provide atomic actions in TMN management systems is far more complex because we deal with networks and equipment as well.

The provision of atomic actions in TMN systems usually deals with:

- operations required at agent systems—CMIS operations to be performed by NEs
- operations required at database, either MIB database or application database—SQL queries or application interface calls
- operations required at GUI, either some internal presentation (topology of network, status of symbols, etc.) or the actual GUI updates (change of color, raise of alarm by flashing icons, etc.)
- operations required to process asynchronous events and the propagation of these events to appropriate managers and applications
- other operations such as event log

With the current insufficient understanding of atomic operations in the network management area and the lack of technology to support them, it is impractical to require the complete support of atomic operations, particularly when the CMIS operations to the network elements are involved.

However, the distributed management system must be designed so that it will support atomic operations to the best effort. In particular,

- all database operations are supported by database transaction functionality
- all operations performed on objects with persistence are included as part of transaction
- all operations with application GUI must support atomic operations
- all asynchronous events during the atomic operation are handled
- all potential failure of atomic operations must be traceable by means of notification or log

In one service provider's billing mediation system, a simplistic design resulted in a system where the call records were dispatched to the billing system through a communication channel. If the communication was broken, the records were directed to a backup billing system through a backup channel. However, since the system was designed without functions to support atomic operations, every time the switch-over happened, it behaved by either sending a duplicated call record, or missing one record depending on the different circumstances of the failure.

5.3.6.3 Availability. To enhance the safety of the management solutions, a high degree of availability of the management system is required. This requirement encompasses the following aspects:

- The service must be available at all times
- The system must have sufficient systems management ability to discover the faults
- The service can be switched to the backup system when faults occur
- The switch to the backup system can be performed automatically and consistently

One vendor SDH management system has demonstrated the need for the system availability. The management system was based on a commercial network management platform which can process about 50 network alarms per second. When the system is deployed in an SDH network, a single fibra cut in the SDH ring can result in over 2,000 alarms per second. Without using techniques such as event correlation (reduce alarms at source) and scalable architecture (increase throughput) to strengthen the management solution, the management system will soon become unavailable.

5.4 MANAGING THE DEVELOPMENT PROCESS

We now understand the issues and complexity related to the exercise of building large-scale distributed systems. Next we focus our attention to the development processes and methodology used to build large distributed systems for our business needs.

5.4.1 The Planning

It is very important that we understand exactly what we are trying to achieve in this exercise. We need to analyze our requirements first. We need to ask why we want to reengineer the existing system. There must be valid business reasons rather than simply applying new technology experiment to the operation systems.

To understand the needs, the list of complexities discussed in the previous sections can be used as a checklist. We need to establish:

- What is the size of the managed environment the software is targeted to?
- What are the system's scalability requirements?
- What are the reliability requirements, having considered the complexity involved to support them?
- What is the overall development road map to determine the extensibility of the system?
- Do we have sufficient understanding of the current level of sophistication and maturity of the key technology?
- What are the system migration and legacy integration requirements with the existing operation systems?
- What is the safety requirements for the overall system integrity and data consistency?
- And above all, what are we allowed to spend?

Depending on the circumstances and the complexity of the system, we may have to allocate resources to prototype in order to discover the answers.

One common difficulty in this process is how to prioritize the requirements. Obviously, we would like to meet all the requirements, the end system exhibits all desired features and the scalability, safety, and reliability will all be provided to the maximum. However, this always contradicts with the amount of resource allowed within the project. Another difficulty is to deal with the key technologies. As in real life, all key technologies, particularly the desirable ones for innovative solutions, exhibit certain level of immaturity. At the planning stage, one obvious task is to minimize the risk by evaluating their functionality and maturity. This can easily lead to the practice of excessive technology evaluation—a common symptom of R&D projects.

To overcome these difficulties, we should focus on the most important activities of the project planning: the development road map.

We should aim for the large distributed object-oriented systems to pay back over time—scalability, flexibility, interoperability, and reuse are all issues that will manifest after the first deployment of the system. At the planning stage, we should consider the following aspects:

- The system will probably be released over several years, and the system will take a few projects to complete—each project and release have to be built with that in mind. The planning stage should have firm plans for the next release, and concepts for the next few.
- The deployed systems will grow with time—end users will want more features, and the system capacity demand will increase. The plan should be made for the software upgrade in mind, the current system is but the first release of a whole software release plan. The system should also be designed with hardware growth in mind—allowing for adding machines, interfaces, and the like.
- Always keep in mind that the system we are going to build will represent a considerable investment—we should exploit that investment by fostering reuse of components within a release, between different releases, and with other systems. We should budget time for engineers to employ reuse techniques such as generic interfaces and meta-data. We should also implement a rigorous documentation process.
- A relevant standard specification from international standards bodies such as OMG that emerges within our problem domain will impact our system—we need to always take the standards strategy as part of the planning process. This includes the different level of participation of different standard bodies:
  —drive the process
  —influence the process

—participate and monitor
—ignore
—try to create de facto standards

- Always assume that the system will be part of a bigger solution, which will be completed in the future—the system should be designed as a building block. This will make adapting to the inevitable business changes easier.

Finally, we should always anticipate that in a few years, we may be awash with CORBA components—we should look at the system as a collection of components, some purchased and some built. It is important to focus carefully on the core components of the system. For example, if the system is a service provisioning system, decompose it into components such as security, user administration, fault monitor and reporting system, and the recursive subnetwork topology management algorithm. We can buy entry level security and user administration components now, we may choose to build a basic fault management module and expect to throw it away sometime later, but our real investment is in the network topology management algorithm—that is our market.

On the other hand, we should also anticipate that in a few years, the market may have favored other technology such as DCOM. We need a strategy if the current plan is to build a CORBA-based system. It needs to be realized that although most development effort goes into the realization of algorithms, the most manifested value of components are their interfaces. We need a strategy to protect these interfaces such as using a more generic technology independent tool to model them. This minimizes the dependency on the IDL specifications, thus it minimizes the impact on the applications when underlying CORBA technology is replaced by DCOM or something else.

The interfaces of a component represent its most important value. This is because these interfaces represent the component's functionality which has global visibility to the overall systems and other components. They represent the value proposition of the component independent from where it is used. When a future system reuses this component, its main justification is the suitability of these interfaces, because the implementation algorithms behind the interfaces can always be improved without any global impact. Also when the systems migrate, the cost is determined by the stability of these interfaces. The cost will rise if these interfaces have to be changed. The suitability and stability of the component interfaces is the fundamental justification and value of standards.

The large-scale distributed systems need considerably more effort to administrate, deploy, and educate the users. These need to be carefully planned as part of the project.

5.4.2 Design Paradigms

Before we design the system, we need to consider carefully the issue of paradigm shift. If we don't have much experience in building large-scale distributed systems and all we can draw is from our experience in building centralized systems, then it is more important that we recognize that there is perhaps a much bigger gap between developing these two different types of systems.

Design paradigms deal with the issues we will face when we build or buy the large-scale distributed systems. If we plan to build or be involved in building these systems, we then need to develop an intuitive understanding of these issues, and their underlying causes.

The most important issue is the building blocks, or components. What constitutes a building block and what is the manifestation of a building block? The classic computer science suggests that a program is algorithm plus data structure. This paradigm defines a building block as a procedure of operations over a well-specified data structure.

In the large distributed systems development, this paradigm has to be shifted to change the emphasis away from algorithms and toward interfaces. Specifically, this change means:

- Interface design must now become the primary architectural output. The later an interface is changed, the more costly it will be. This is evident that in a distributed system, the interface specification manifests the functionality of the component. This interface should always outlive the technology used to implement the component.
- Algorithm design takes a multiphase approach:
  —quickly build a rough implementation of all services with key functions only, which are nondistributed, handle no corner cases, and are slow
  —publish these to free up the development of fully functional but nonoptimized services
  —profile the system's performance and fix bottlenecks
  —defer better implementations to later releases—at that time, the operational characteristics will be better understood.
- Where it affects interfaces, support the notion of generalization and specialization in the design from the start. If we have a problem X, we always try to generalize it into a form Y, with X being a configuration (a specialization) of Y. This notion is the abstraction principle of the object orientation and it has a pivotal role in the development of large distributed systems. It is fundamental in building semantic abstraction, enhancing reuse and technology migration. Some later chapters further discuss this issue in the area of object modeling and technology integration.
- Data management issues such as storage technology are hidden behind interfaces, therefore, data representation moves from being a strategic decision, as it is in an RDBMS-centric system, to a tactical decision. A standard object-oriented database management system can be a good solution to support distributed objects. The key is that the data management becomes a local issue and their role is to support the component's interfaces.

When we design the large distributed systems, we need to consider the following interfaces:

- *Downward*—Clearly we will use third party components and, therefore, we need to conform to these interfaces. However, how much do we trust the stability of these interfaces?
- *Sideways*—We will have several subteams building the subsystems. History's lesson is that defining subsystem interfaces early in the project and tightly controlling their change is a prerequisite for project success.
- *Upward*—Never consider the system to be the end of a business process. Always consider it part of a bigger picture. This needs to be reflected in the design. The corporate world is littered with systems that come to a dead end with a UI—operators read off one screen and type into another (or the GUI equivalent). In today's software jungle, the survivors are the team players, rather than the stars.

Large distributed systems often grow out of single-host systems. It overcomes the scalability problems suffered by the small systems. However, the users of these systems often expect the new distributed systems to behave just like the old ones, only bigger. This is not achievable, we cannot make a big system look like a small system to the users. For instance, the user interfaces may define global operations such as:

- *global searching*—operations that display all *X*'s that conform to condition *Y*

- *global naming*—users expect to name things with names like *Mary* or *Subrack 6*
- *global access*—users can operate on any object in the system

These global operations are undesirable in a large distributed system—the globe is a large place. It is hard for a designer to get a feeling for *global* because a system deploying 100 times larger may have only twice the code.

Large-scale distributed system users should never ask *global* questions. If, on the rare occasion users do need global information, they must assume that such queries are slow and costly.

The mind set is that on the single-host system, the global queries were actually scoped by the machine. For example, *display all Xs that conform to condition Y* (with an implied condition that X exists on this machine).

A query interface design principle may need to separate the concepts of global transactions and global queries. While the global queries are undesirable in some applications, the global transactions may offer the benefits. The global transactions rely on the ability of the components in the distributed system (either close or distant) to *process* very complex tasks and to handle data management issues. The separation and careful use of these two concepts allow each to evolve and be optimized accordingly to suit the application needs.

Reflecting on the interface design, we should always:

- Design for differential service—*close* is fast and *distant* is slow
- Introduce an object structuring system such as domains
- Introduce fully distinguished names
- Introduce a security concept

More importantly, this should not only be the paradigm shift required by the software designers, this is also the paradigm shift required by the users of the system as well. So all these should be exposed to the users as part of the end-user education for large-scale distributed systems.

5.5 TECHNOLOGY MIGRATION

There are many distributed object computing technologies which can be used to implement the processes discussed in this chapter:

- *CORBA*—Object Management Group's Common Object Request Broker Architecture, which is becoming a de facto industrial strength distributed object oriented technology for a wide range of application areas.
- *Open Distributed Processing*—ITU-T and ISO have defined a framework of using viewpoints from which to abstract or view distributed systems. A set of concepts, structures, and rules is given for each of the viewpoints, providing a *language* for specifying open distributed systems.

  ODP defines the following five viewpoints:

  —**Enterprise viewpoint**—focuses on the purpose, scope, and policies for the system.

  —**Information viewpoint**—focuses on the semantics of information and information processing required.

  —**Computational viewpoint**—enables distribution through on the functional decomposition of the system into objects which interact at interfaces, e.g., objects, interactions, bindings, and binding types.

—**Engineering viewpoint**—focuses on the infrastructure and mechanism required to support distributed interactions between objects in the system, e.g., clusters, capsules, nodes and channels.

—**Technology viewpoint**—focuses on the implementation technologies.

- *TINA-C*—TINA Consortium has defined an integrated architecture for telecommunication management applications. The base computer systems architecture is the *distributed processing environment* (*DPE*) [96]. DPE is a distributed computing platform with special features to support telecommunication management functions. The features include multiple interfaces in its *object definition language* (*ODL*), the incorporation of the *general relationship model* (*GRM*) for defining relationships between managed objects. This distributed environment provides the base architecture to support TINA-C's service architecture and network resource architecture.
- *DCOM*—Microsoft's proprietary distributed object computing technology, widely used as a platform to integrate a large range of Microsoft applications.
- *Other middleware technology*—There are DCE RPC-based middleware, secure message-based technology, and on-line transaction processing technology which can be used as general purpose distributed computing platforms.

There also exist the specific instances of the distributed technology applied to the TMN network and service management. In addition to the TINA-C's distributed processing environment which is specially created for management telecommunication networks and services, the OMG Telecom Task Force also defined a telecommunication network management framework based on CORBA. In ITU-T, the work is progressing to apply the principles of the open distributed processing to the OSI reference model to create the *Open Distributed Management Architecture* (*ODMA*) [81].

In the past decade, CORBA has emerged as the most commonly used and commercially advanced distributed object-oriented technology in very diversified industry domains. This also applies to the telecom industry. Although TINA-C network and service architectures offer considerably more functionality in telecom management, most of its global demonstration showcases and R&D efforts are developed by using CORBA technology as the underlying infrastructure support. CORBA is also used as a technology to support ODP's engineering and technology viewpoints. TeleManagement Forum's Technology Integration Map also firmly endorses CORBA as the base technology to integrate multiple TMN management functions and applications. Some large-scale commercial system and application developments using CORBA are also underway, and the commercial ORBs and CORBA services are available with reasonable strength and flexibility.

In the TMN development, CORBA's value-add to TMN management applications include:

- It offers a general purpose distributed computing environment.
- It offers high-level interface description techniques to specify the function of management applications and the interface property of managed resources.
- It offers an integration platform for different management functions, applications, and systems (not necessarily developed in CORBA).

CORBA achieves these by its three major architectural components: an object request broker, an IDL language, and business object definition facilities, and a set of CORBA distributed services.

In the remainder of this chapter, we briefly discuss the characteristics of CORBA in the context of developing TMN management applications. Readers can refer to some literature included in the reference section for more detailed technology features.

5.5.1 CORBA as an Application Integration Platform

The concept of CORBA as an *integration bus* to integrate various entities in an application is one of the main benefits of using the technology. Generally speaking, people use CORBA to develop some applications and components with specified functionality. These applications and components need to interact with the resources in a domain specific environment. These resources include databases (relational or not), legacy systems, operation support systems, networks with hardware, customer access, and presentation devices, etc. Further, these applications need to interact. The use of CORBA can reduce the task of facilitating interactions between these entities, and allow developers to concentrate on developing components.

The interface definition language IDL is used to describe the object interfaces between the components that interwork through ORB. This interface design is crucial to the success of the CORBA-based development. Recall the discussions in Section 5.4.2, this interface specification should be regarded as the most important architectural outcome and be made stable as early as possible. One should notice that IDL is only an interface specification language rather than a language that can be used to specify systems, components, and interaction behaviors. In a large distributed development, higher level specification tools (object-oriented design and analysis tool) are usually required to focus on the enterprise and business specifications before the implementation is considered. Usually, these tools provide translation to IDL that can be used for systems engineering. In one of the later chapters, we discuss the principles and processes of object design and specification without targeting a specific language.

One of the key benefits of using CORBA as an integration bus is the ability to wrap legacy systems or external systems with IDL interfaces. This is a common technique to make CORBA-based applications interworking with legacy or foreign systems. Again the key issue is the business specification of the wrapper—it defines the business level view of the intended behavior of the legacy or external systems (systems we don't have any control over such as corporate databases controlled by other departments). This specification is a strategic architectural outcome, and it is independent from the behavior of the individual systems they are representing. The implementation of this interface to individual legacy systems is a tactic task—the mappings can be changed at any time when the external systems evolve. When the corporate database migrated from relational database to object-oriented database, we don't want to rewrite the software which uses the database. We only change the implementation of the wrappers.

Figure 5.4 illustrates the concept of the CORBA integration bus. In the scenario, a number of CORBA object building blocks interact with each other by using ORB. These applications also interwork with some external systems through the use of wrappers.

5.5.2 CORBA Service

A large number of *Common Object Service Specifications* (*COSS*) [89] have been defined by OMG and some services have been implemented by commercial CORBA vendors. In this section, we only briefly introduce those services which play important roles in the context of TMN management. Large quantities of literature included in the reference section provide a more complete view of CORBA services.

5.5.2.1 Transaction Service. OMG has published an Object Transaction Service (OTS) which incorporates the architectural features of OSI/TP [76] and X/Open TP technology [101]. It has defined an object-oriented interface to support two phase commit semantics, recovery, synchronization, consistency, and nested transactions. The commercial products

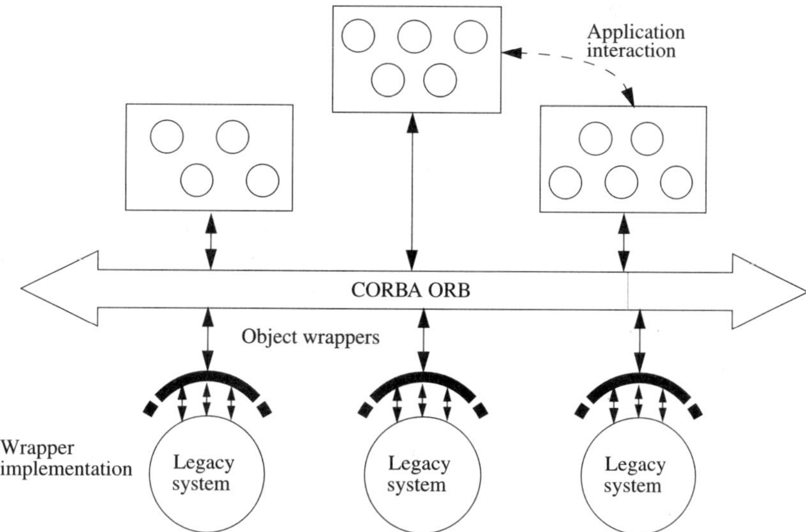

Figure 5.4 CORBA as integration bus.

exist which implement the OTS by utilizing the traditional TP monitors such as CICS, Tuxedo, and Encina.

The ever-increasing demands for distributed management functions have greatly increased the complexity of management applications. Nowadays, these applications have to manage a large number of heterogeneous systems distributed in a global network of computer and communication resources. The heterogeneous nature and global distribution of these resources create new challenges and requirements for management technology. One of the challenges is to provide reliable, consistent, and robust services to coordinate the updating of these resources. DTP technology provides a functionality to meet this requirement.

The DPT technology has found its way into many application areas, including TMN solution development. For instance, a standard for providing OSI systems management in a TP application context has been defined by the ISO OSI committee. Within the DTP application context, the use of DTP facilities by systems and network management functions will further enhance the capabilities of the standard. This combination of systems management and DTP technology enables application developers to build more powerful distributed TMN applications.

The environment targeted by the OSI network management services is often widely distributed. There are occasions when the requested operations and changes to the application data must be made to different systems in a coordinated manner, i.e., the changes must be:

- all performed, or,
- none performed.

In the real-world distributed environment, it is obvious that the requested changes or operations to the distributed managed systems may not be easily coordinated due to reasons such as:

- A failure in some managed system or communication link will prevent the atomicity of the changes or action activities from being guaranteed.
- The change or action is invalid in a certain part of the distributed management system.
- The shared resource in the distributed system is (temporarily) unavailable or becomes unavailable.

- The operations requested cannot be carried out due to security and access control requirements.

Experience has shown that a very significant part of any application design, production, and operation effort is spent on failure recovery and error handling in order to safeguard the integrity of the application data. Partly performed operations are the main source of inconsistency and should be avoided.

In one of our commercial R&D development, we built a path management system for managing broadband networks. The network was designed as a hierarchical network with nested subnetwork architecture. The path management function was delegated recursively following the subnetwork structure. Since each subnetwork is managed by an instance of the path management system in a separate location, the path operations (create and delete) are truly distributed operations.

A major design complexity was to make sure that if the portion of delegated path operation failed on one subnetwork and eventually caused the failure of the whole operation, the other subnetworks knew about this failure and kept their data consistent. This proved to be a very difficult to design and when the solution was deployed, its throughput was limited by the fact that a large portion of system resources (networks, nodes, termination points, etc.) had to be locked to safeguard the data integrity.

In the second phase of the project, a commercial transaction monitor software was used to provide a generic solution for failure recovery and data consistency. This shifted many design complexities to the transaction monitor thus considerably simplifying the application development. Although the round-trip time for individual operation didn't improve, the overall system throughput was increased by 30% as more concurrent operations can be executed.

A reliable and reusable solution is required to coordinate the management operations in a distributed management system. The solution should be designed in such a way that:

- update consistency is maintained throughout the activity;
- exception handling is provided in such a manner that consistency is maintained and robust solutions are provided;
- the solution is generic and independent from the real-user application semantics such that it is reusable elsewhere or by other applications;
- the solution is *open* so that off-the-shelf packages can be purchased and application portability and interoperability can be achieved.

The design of the OMG object transaction service attempts to provide such a solution. It promises to remove, from the user application, much of the burden of coordinating consistent changes or actions and of handling exception conditions such as communications failures and system failures to give a consistent result.

Figure 5.5 depicts a scenario where OMG object transaction service is used to coordinate application access to multiple resources. These resources can be network management systems, OSS, or enterprise information systems. They can be CORBA or non-CORBA components.

5.5.2.2 Trading Service. Trading Service published by OMG is a useful tool in large distributed environments because they release users from the growing problem of finding objects which can satisfy a set of prescribed properties. The OMG trading service allows object service providers to advertise their services by publishing the service features, and allows object service users to search for services. Trading service acts as a matchmaker for both consumers and producers. The implementations of trading service, commonly referred to as Trader, are available as commercial products.

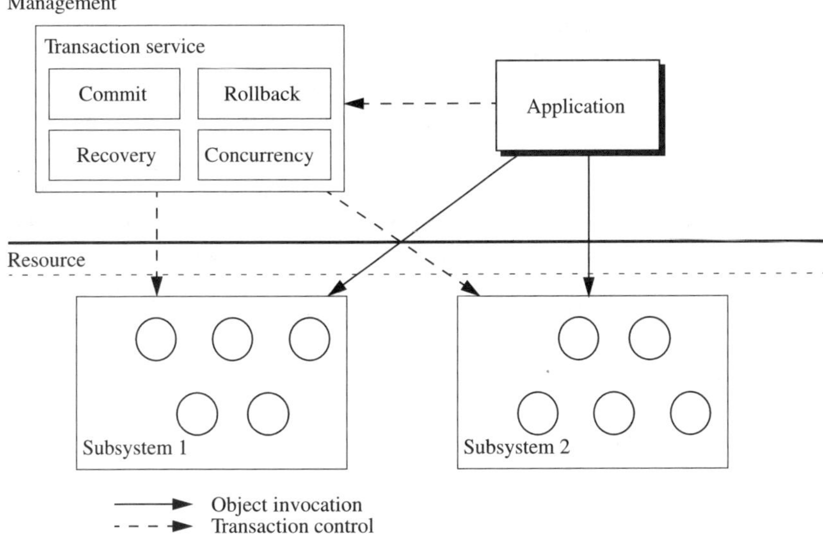

Figure 5.5 The use of transaction service to coordinate multiple resources.

The potential of using ORBs over the enterprise has already demonstrated the simplicity of trader because it is an intelligent means of locating an appropriate service from literally thousands of servers and then automating the services advertised quite transparently to users.

In the large networks environment, through dynamic discovery of object properties, searching becomes largely automated and traditional tasks such as maintaining links to information, objects, and network connections are no longer required. Users automatically find updated information as it becomes valid through advertising and discovery of objects with the appropriate characteristics.

For example, the properties of application logic and business rules inherited in a network or service management application can be kept as objects and are invoked by system procedures dynamically. Updating to these logic and properties requires only the object properties to be updated once and all applications are given the new information when they next enquire. This will greatly increase the consistency of the information in the distributed environment.

In a customer service management scenario, a service provider needs a service and product catalog as part of the sales process. This catalog can be managed by a federated trader in a distributed enterprise environment. New services and products can be added to the catalog, and customers can search the catalog for desired services and products by providing descriptions of them. Figure 5.6 depicts this scenario.

5.5.2.3 Relationship Service. OMG Object Relationship Service facilitates the management of relationships between CORBA object instances. In the CORBA model, the relationships an object is participating in are managed separately by the relationship service.

One of the most useful functions of the relationship service in TMN context is to provide object mapping between different TMN management hierarchies. Figure 5.7 depicts a scenario where relationship service is used to maintain the information of object mappings. Three types of mappings are required in the application:

- *Network element mapping*—A service object, represented by an SMO, is implemented on top of a set of network entities which are modeled as network management objects (NMOs) which are in turn mapped into a layer of network elements.
- *TMN management object mapping*—A mapping capturing the relationship between

these two sets of objects needs to be maintained by the service applications. These relationships are maintained independently of the object implementation. That is, an object should not be affected by what relationships it participates in, nor should it be affected by the changes to these relationships. This provides flexibility to do things like changing the network components which provide a service without affecting the service instances.

- *TMN management function mapping*—An operation on a service object may result in changes in the corresponding network objects which reside in the network elements. Therefore, when an operation is carried out on the service object, the service application needs to map it to the appropriate operations of the corresponding network objects. On the other hand, a set of services may be affected when a fault (event) occurs in one of the network elements. This event needs to be propagated to the corresponding service objects by the management system.

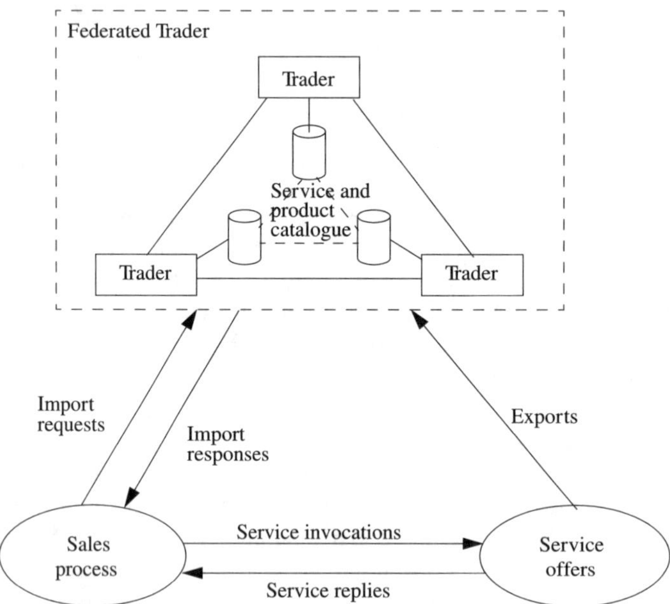

Figure 5.6 Federated trader as service and product catalog.

Relationship service provides the management of these relationships independent from the object semantics. The object semantics can be enriched by adding more relationships in an incremental development manner.

5.5.2.4 Notification Service. The OMG Event Service offers a limited amount of functionality for the propagation of events, including direct transmission between producer and consumer, and indirect transmission via an event channel.

The CORBA Notification Service is an extension of the event service targeting specifically the telecommunication applications. It offers an extra level of indirection; events are produced and consumed using configurable event channels (CORBA objects that are always accessible on a local server). The notification service is responsible for distributing events from producers to interested consumers.

Figure 5.8 depicts a scenario of using CORBA notification service to propagate events generated by one producer (e.g., a CORBA agent managing an SDH equipment reporting a fiber cut) to two consumers (e.g., connection manager, and fault manager) which are interested in receiving these events. The producer and consumers can be located at different hosts.

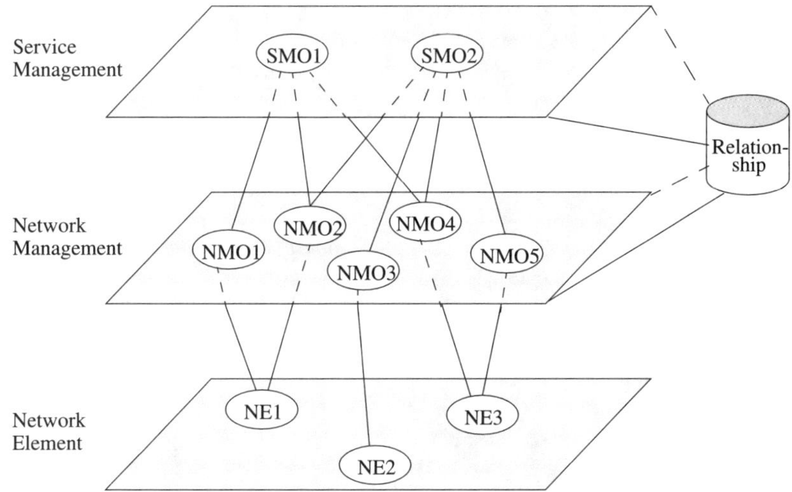

Figure 5.7 Relationship mapping between TMN layers.

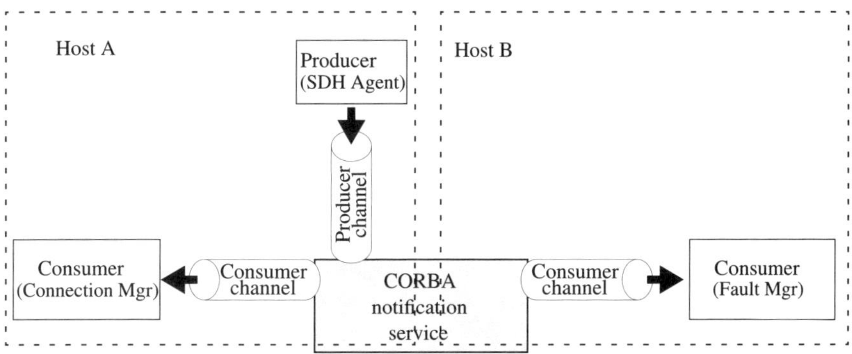

Figure 5.8 Scenario of using CORBA notification service.

5.6 FROM TECHNOLOGIES TO SOLUTIONS

Migrating to distributed systems requires many efforts: understanding the requirements of the migration; understanding the difficulties and impact of employing distributed technology; managing the processes and paradigm shifts; and understanding of the use of technologies in a TMN solution context.

Above all we must understand that the purpose of the technology migration to distributed systems is to build distributed management applications. The telco business process reengineering is a process of building management solutions which support more processes and applications to manage customers, services, and resources in an integrated and scalable manner. The distributed object technology is used as an enabler to achieve these business goals.

The challenge is learning how to use these distributed technologies more effectively to achieve the business goals. In the next chapters, we discuss the issues of building distributed management applications using distributed object technology to model, architect, and design the integrated management solutions.

6

Building CORBA-Based TMN Applications

In the previous chapter, we discussed the issues related to building large-scale distributed systems and the technology advancement in this area. In this chapter, we examine the applications of these technologies to the TMN management functions and systems. In particular, we examine the distributed OO technology (as represented by CORBA) and its applications in more traditional OSI-based TMN management. We also discuss the issue of integration between these technologies.

The focus of this chapter is to position CORBA in the TMN development. The issues examined in this chapter include:

- Are there any requirements for distributed OO technology such as CORBA in TMN development?
- What are the technology characteristics of TMN network and service management system?
- Exactly how can CORBA add value to TMN development?
- Is CORBA a replacement technology for OSI which is so far the main focus of TMN development?

As we observed before, TMN is a management framework and its principles and functions can be implemented by using different technologies. Our experience has shown that CORBA, when used properly to address certain class of problems, can add great value to the TMN development particularly in the distributed network and service management areas. CORBA is not simply a replacement technology for OSI; it should be seen as a great value-add and its integration with OSI delivers considerable benefits to TMN application development.

In a case study at the end of the chapter, we present a distributed network management system implemented according to these principles. The architecture of the system uses CORBA and its integration with OSI CMIP to deliver distributed connection management functions in a hierarchical and heterogeneous network.

6.1 IMPLEMENTING DISTRIBUTED TMN APPLICATIONS

The business process reengineering prompted by the global competition and deregulation in the telco industry has intensified the need to build a service management environment, where the telco service offering environment, service creation environment, service provisioning environment, service management environment, and service deployment (based on the network) environment are all integrated in a single architecture.

In this chapter, we address the technical challenges of using distributed technologies to build an integrated service management environment. We discuss the value-add of these technologies to TMN management. We also discuss some technical issues specific to the TMN network management. Finally, we present an example of how these technologies can be used to build integrated TMN network management systems.

Derived from the business objectives and requirements, the technical challenges for the integrated management environment include:

- *Distribution*—Telecommunication services are geographically distributed. Hence, the management applications should also be distributed. This requires the management platform to support highly distributed management applications by providing distributed environment and distributed services.
- *Scalability*—Ever-increasing size of users, resources, information, services, and networks to be managed has required the management platform to be highly scalable.
- *Heterogeneity*—The extremely diversified types of telecom services, resources, customer care requirements, and computer systems in the management domain require the management platform to handle heterogeneous systems.
- *Common services*—Some distributed services are common to many distributed applications. A platform which supports these common services will greatly reduce the cost and complexity of application development.
- *Component-based development*—The monolithic software development and deployment practice in the past is one of the main causes of the legacy systems and integration nightmares. The new environment must support component-based development of functionality to provide a better chance for the software solutions to be future friendly.
- *Consistency*—Data consistency is a very important issue in distributed telecommunication network and service management applications. The management platform needs to have the ability to support a consistent data model.
- *Reliability and performance*—A high degree of reliability, availability (7×24 is a common requirement), and performance is crucial to the network and service management applications. This is particularly important when the size of networks and complexity of services increase dramatically.
- *Service integration support*—The support for service integration is a new issue for management platforms. Integrated service management functions are required so new services can be easily introduced, deployed and managed, and integrated with the existing services.
- *Interoperability*—The management platform must interoperate with different business and administrative domains. These domains may have very different enterprise policies, business focuses, underlying technological frameworks, and organizational infrastructures.

All these issues reflect the complicated requirements to build distributed and integrated TMN management system applications. It is clear that migrating to distributed object-oriented technology framework is essential to meet these technical challenges.

Technically, distributed TMN management systems share much commonality with the distributed computing discussed in the previous chapter. They manage large distributed environments, and the principles and approaches of building distributed systems apply to building TMN systems. However, there are many different characteristics in TMN systems. These characteristics pose new challenges to the application of distributed technologies to TMN.

There are three distinct environments in a typical TMN system. These environments deal with different telecommunication business needs:

- TMN systems support services. The entities representing services need to be modeled.
- TMN systems support *managed* environment such as networks and elements. These entities and their functions directly support services. The management of this environment is unique compared with the general distributed computing functions. These managed entities need to be modeled in the system, and traditionally, they are modeled differently from services.
- In addition to managing these two separate computing environments, TMN systems also support management activities which ensure that the services are supported well by network and element functions. These management activities are represented by entities which need to be modeled in the system.

To build an integrated TMN system is not only to support these entities and functions. A high-quality TMN system also requires such technology support that these different aspects of the system can be all integrated seamlessly to achieve the service satisfaction by the customers, smooth operation of the network, and the management of their interactions.

The crucial question then is exactly what values the distributed object-oriented technology such as CORBA can add to the TMN framework to achieve the business goals and technical outcomes. Before answering this question, let's first look at the different characteristics of the TMN network and service management functions in more detail.

6.2 CHARACTERISTICS OF NETWORK MANAGEMENT SYSTEMS

TMN network and element management is responsible for the smooth operation of telecommunication equipment. TMN network and elements are normally modeled by a large number of relatively simple objects. These objects and their relationships model the real physical network elements, network topology, and network hierarchy. These entities are relatively simple in the sense that objects normally do not have complicated operations and relationships. The system behavior and possible events associated with them can be well defined. Most operations performed on these objects are simple and can be performed on a group of objects determined by a specification of object properties.

TMN network management is a set of activities which can be performed on managed objects representing networks and elements in order to configure the network structure; monitor the status, usage, and performance of networks and elements; control the behavior of these network resource; and detect and repair faults in the network, etc.

Normally, a network management system contains the following major components:

- A number of management entities which play manager roles. These management entities have global views of network resources within their portion of the management hierarchy. They are driven by management applications or directly accessed by telco operators to issue management requests on their behalf. A manager normally communicates with a number of agents. A manager process is normally located at a central office, rather than on each managed equipment.

- A number of management entities which play agent roles. These management entities have better views of the network resources they are assigned to manage, such as the status of network elements, the usage data, and the connections between them. They are responsible for all aspects of managed objects, the storage and retrieval, operational semantics, object states, events, etc. Compared with a manager, an agent has a more focused view on a section of a managed network. An agent process is normally located close to the network resources they are managing. Sometimes an agent process is built on the equipment it manages.
- A communication platform which is used between the managers and agents to send and receive management information. A communication protocol is used between managers and agents. OSI CMIP is adopted by TMN as the network management protocol.

Telecommunication network, once established, is relatively stable. Network elements often have physical locations associated with them. For example, a switch is installed in a local exchange and it has a number of ports. A network management system's main task is to manage the physical network resources. This task has the following characteristics:

- The status of the network resources, as represented by the network objects, is relatively static.
- The hierarchical network architecture and commonality among network elements (objects) often leads to group operations on managed objects.
- Agent processes are normally close to managed network elements while manager processes are normally in central office. This leads to centralized management.
- The distribution requirement is mostly on agents, not so much on managers.
- The scalability is required only when the network resources grow.

As indicated before, the OSI management technology was designed within the context of these characteristics, as such, OSI is chosen to support TMN element and network management.

6.3 CHARACTERISTICS OF SERVICE MANAGEMENT SYSTEMS

Compared with the network concepts, TMN service concepts are normally modeled by some more complex objects. These objects represent high-level abstractions of the network functions and the value-add to the customers. They also represent the relationships between network components (network elements) and computerized services supported by the network. For instance, a teleconference service may have a number of instances representing the distributed service offerings and management components. These components may include a number of end-to-end broadband connections carrying teleconference contents with a certain quality of service requirements. Further, this service is supported by a large number of underlying network elements and various management information systems. In some cases, hundreds of thousands of objects may be involved in setting up end-to-end broadband connections. Access to a teleconference service may invoke very complicated functions including accounting, security, performance, interoperability, quality of services, etc.

TMN service management is to manage telecommunication services through the management of service objects interacting with the network resources.

One of the major differences between network and service management is that services often need to be created, implemented, and deployed dynamically in real time while networks are relatively stable by comparison. Service management often has the following characteristics:

- Service objects are more dynamic, as they represent the life cycle of shorter but more real-time service transactions.
- Service objects are more location independent. A service object may start its life cycle from a sales representatives laptop and may follow it through to the networks managed by the service provider's partners.
- As every service provider is aiming to provide more services with the existing network infrastructure, high scalability is required and this scalability is not necessarily limited to the network scalability.
- No clear management technology directions have been set within TMN framework as to how TMN service offers and service management functions should be implemented.
- More interactions with the enterprise information systems are required, which resembles more of an enterprise IT management environment.
- Both service offerings and service management functions are distributed. The distribution model may be governed by different policies and strategies.

Although TMN did not mandate any specific technology to address the service management needs as it did in the network management, it is clear that considerable change of characteristics in service management requires the new management technologies to be introduced.

6.4 CORBA FOR TMN MANAGEMENT

Given the characteristics between the TMN element/network management and the service management, it is obvious that the technology requirement for the former focuses on *open,* whereas the technology requirement for the latter focuses on *distributed.* OSI technology was designed as an open technology whose main feature is to facilitate open protocol communications between two application entities. Although some additional features of the technology were introduced (such as directory service to provide location transparency), the main focus of the technology is not on providing distributed application development environment. This reflects its focus on the requirements to provide a technology framework for element level interoperability.

CORBA technology, on the other hand, was designed to support distributed applications from its inception. It supports a general purpose distributed object-oriented paradigm. It is focused on providing object-oriented distributed application development support and object level distribution. Table 6.1 illustrates the main differences between OSI as an open and CORBA as a distributed technology.

Given the technology requirements for TMN management functions, there are two major areas which CORBA can add value:

- CORBA can bring its distributed functionality, distributed object model, and a whole range of object services to assist the development of distributed network management applications.
- TMN service management is a layer based on the network infrastructure, and its object modeling poses different requirements from the network object modeling. CORBA technology can add more value to the TMN service management domain.

The value CORBA can add in the network management area is its distribution capability and generic object services. CORBA's distribution mechanism plays a number of major roles in developing large-scale distributed network applications:

- It can be used to distribute management functions.
- It can be used to construct distributed management applications based on the management functions.
- It can be used to distribute management domain (OSI networks, databases and other resources).

TABLE 6.1 OSI and CORBA Technology Features

Environment	Open	Distributed
Distribution unit	Management entity	Object
Naming	Management entity + RDN	IOR
Location	Nontransparent	Transparent
Target application	Management of telecommunication elements and networks	General purpose distributed systems

Figure 6.1 depicts this architecture with CORBA providing the distribution framework for management applications.

In this architecture, CORBA is used to develop network management applications or components. For instance, the management of a subnetwork in an ATM domain can be realized by a set of CORBA objects, each representing certain network management functions such as connection management (create connections, delete connections, and report connection status), and network topology management (such as create subnetworks, create links between subnetworks, and remove links). Section 6.6 presents an example of using CORBA to create a path management application for a standard ITU-T G.805-based broadband network architecture.

Another area CORBA can add value to the TMN environment is to build a set of object adapters which map abstract CORBA requests to network management operations. This approach utilizes the key feature of the object-oriented paradigm in CORBA to build CORBA interfaces to the existing network element functions, databases, and other OSS functions. This allows the integrated management between CORBA applications and the under-

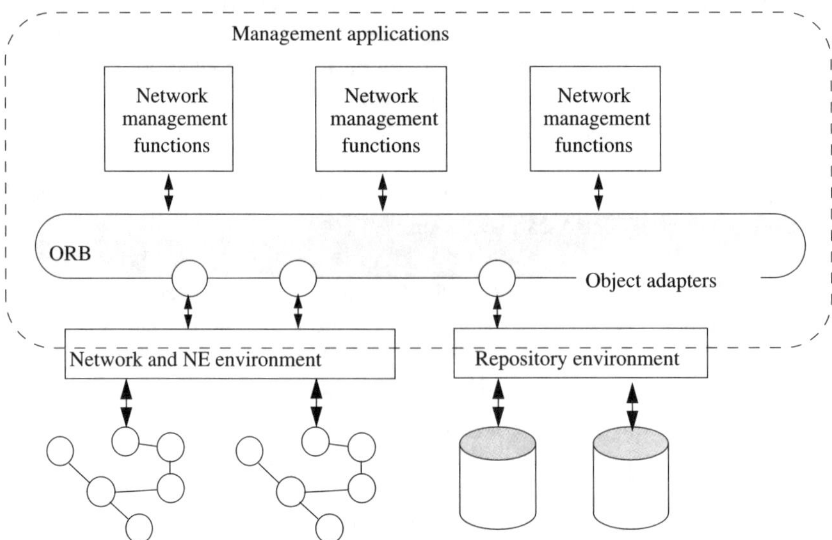

Figure 6.1 Distributed network management architecture based on CORBA.

lying non–CORBA environment. The CORBA/OSI integration technology plays an important role in supporting this linkage. This is discussed in more detail in the later chapters.

CORBA can also be used to provide the specification and implementation for a layer of higher level objects which represent the cross network technology domain functions. Figure 6.2 depicts a scenario that CORBA is used for cross-domain connection management.

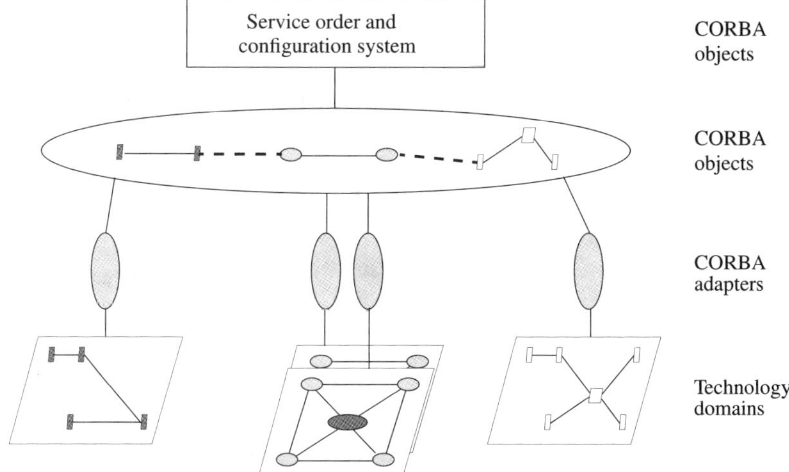

Figure 6.2 CORBA for cross-domain network management function.

In this scenario, the business applications require the underlying network management system to provide an end-to-end, integrated view of the network function. This view is independent from the current network technologies. In a typical network deployment, these technologies have their own management systems with different interface specifications. For instance, an ATM subnetwork may present a standard ATMF M4 public network view for its connections whereas an ADSL equipment may present a nonstandard view for its connections.

A unified end-to-end network interface can be defined by using CORBA IDL, and the management system of this interface can be developed by using CORBA. A number of adapters can be produced to adapt the unified interface to the individual technology domain. In this way the business applications will be able to use this CORBA interface to access network functions in a generic and end-to-end manner.

In the TMN service management layer, TMN services are difficult to model due to its dynamic features. CORBA technology offers a better alternative to model these complicated service objects for TMN service management.

CORBA provides facilities for the definition and implementation of objects. It also provides an environment for the invocation of objects. The CORBA IDL allows complicated operations to be specified with generic interfaces. The actual implementation of an interface is totally independent from its definition. This enables service objects to be rapidly developed and deployed.

These features make CORBA technology an ideal candidate as an environment to implement TMN distributed services management. In particular, this environment can be based on the distributed network management environment presented as in Figure 6.1. A building block approach can be used to add more functionality to form an integrated service management environment. Figure 6.3 depicts this network and service management architecture.

The CORBA-based network management environment can be easily extended to support services. In Figure 6.3, the service management function blocks do not represent support for an extra layer of management platform; they only represent an extra layer of object

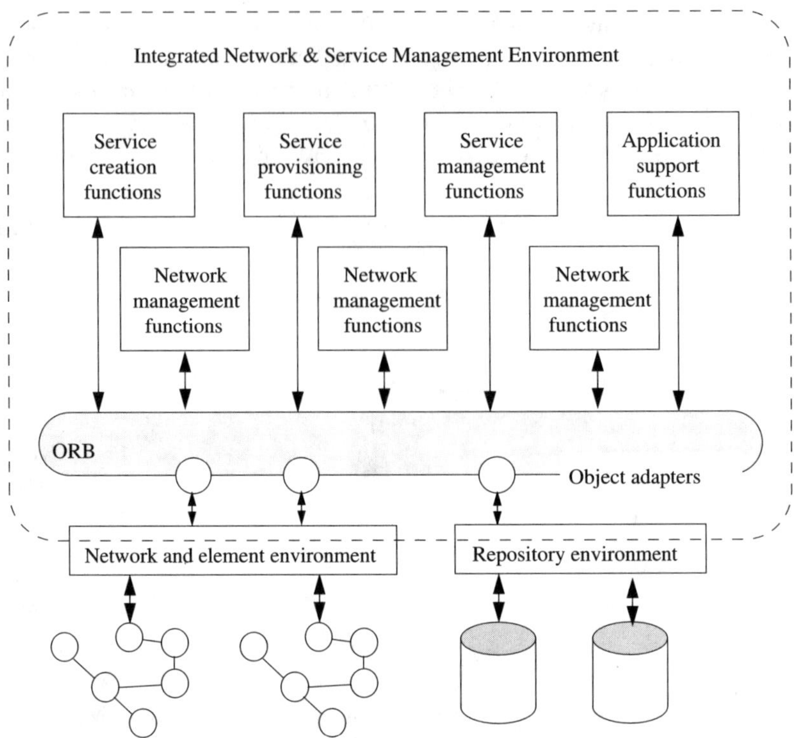

Figure 6.3 Distributed service management architecture based on CORBA.

definitions and their implementations on top of the network management interfaces. Using the cross-domain example, the service creation and management objects such as video-conference service and its management can be represented by a group of CORBA objects invoking cross-domain network management functions (create path, report status, etc.) through the same ORB. This is depicted in Figure 6.4.

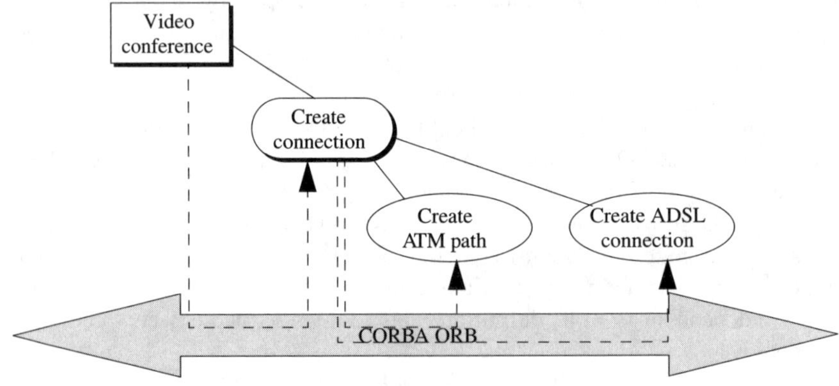

Figure 6.4 CORBA for service creation and management.

TINA-C has also identified the limitations of TMN network management technology for service management. In its service architecture, the services and their management functions are firmly supported by a distributed object technology similar to CORBA. The Distributed Processing Environment is defined as the standard environment for providing distributed communication infrastructure between TINA compliant applications, and interoperability support to non-TINA compliant applications. This technology is currently demonstrated worldwide by many TINA-C members using existing CORBA implementations.

In TeleManagement Forum's Technology Integration Map [100], CORBA is also used as a core integration technology to support distribution, application development, TMN vertical integration (integration between service and network management), and horizontal integration (integration between service management functions and service management domains).

6.5 CORBA AND OSI

The conclusion that CORBA can add considerable value to TMN makes one wonder whether CORBA is a replacement for OSI. Why don't we just use CORBA to implement all TMN layers including element management and network management?

Indeed, many people are asking these questions and many are experimenting with the idea by implementing management agents using CORBA. However, we need some more thinking and analysis to answer these questions.

First, what is the reason we want to replace OSI technology? One of the most common criticism (perhaps the most important one) of TMN and OSI is its complexity. But exactly what aspect or component of TMN/OSI technology is complex? Which part of it needs to be replaced? Can CORBA do better in those areas?

Second, how do we utilize CORBA technology in TMN? Do we reimplement the OSI model or we try to encapsulate it by adding more value?

Using OSI technology to develop management applications is complex. However, many people's criticism of the complexity is focused on the OSI's computational architecture, i.e., the communication protocol stacks. In reality, this is not the main reason OSI is complex. The computational architecture of OSI deals with the definition of application components (management entities), the object modeling approaches, management functions, the way to exchange information between two entities, and the layered architecture to support this communication. In hindsight, this architecture may be designed marginally better, but there is no intrinsic complexity within this architecture. Moving from a centralized software development to an open (somewhat distributed) environment, people need a great deal of adjustment of a complex architecture to deal with complex requirements. In particular, most of the architectural complexity is now hidden by the commercial platform which is constantly evolving to present a much simpler interface to the application developer. The analogy can be found in CORBA architecture itself that, as an application development platform, it has presented a much higher level interface by hiding all details of the RPC paradigm which is by no means simpler than OSI CMIP and CMIS.

The complexity of the TMN development using OSI technology is caused by the complicated object specifications of the telecommunication resources: the elements (switching and transmission equipment) and networks (with different technologies). Most of the MIBs defined by ITU-T are very complex. An analogy is that the contents of parcels handled by a post office are a number of magnitude more complex than the way the post office handles its parcels. The complexity presented to the TMN application developers is that they have to implement every detail of the object model: design each object, design each management entity to represent a portion of management hierarchy, code every method to implement desired behavior, etc. In comparison, selecting a software platform to conduct CMIP communication is relatively much simpler.

Many telco standards bodies, such as ITU-T, ETSI, ATM Forum, etc., have defined a large amount of GDMO object definitions for different types of networks and network elements. Some specifications are based on more technology independent network architectures and object models, and some are more switching and transmission technology specific. Also, some specifications define capability for element management and others define the network or subnetwork view of the management capabilities for the purpose of network management at a higher level of hierarchy and abstraction.

These specifications are needed to implement the element and network management functions. They are essential to the operation, administration, and management of the smooth running of the telecom resources and networks. They are complex, but probably are no more than just reflecting the complex nature of the telecommunication network environment.

Only by replacing the large library of object specifications can we simplify the TMN application development. Otherwise, criticizing the OSI computational architecture is blaming messengers for the bad news. It has been widely reported in literature that using CORBA to implement these complex object specifications do not necessarily reduce the application development complexity. In the TMN development world so far, many efforts have been made into the development of OSI management platforms and agent development tools, which can be used to generate template managers and agents from GDMO definitions. Due to lack of these tools, it will be more complex to implement these specifications using CORBA. In a purely software engineering project management sense, it is extremely hard to recruit software engineers who are fluent in both GDMO and IDL.

In other words, although CORBA is a desirable candidate to implement TMN network and services management components, it is not the best choice of communication protocol for TMN network and element management for the following reasons:

- In TMN network management, network topology and distribution of resources are relatively stable, and the feature of transparent handling of distribution of objects provided by CORBA will in many cases represent an overhead instead of a benefit.
- The OSI manager–agent communication model is well suited to managing a large quantity of network and element objects which are not highly distributed and dynamically reconfigurable. The managers and agents using established peer-to-peer communication channels are more effective than those using a request broker.
- Given the network object hierarchy, it is a very ineffective use of CORBA to model the OSI filtered and scoped operations and containment structure.

Finally, replacing a large segment of established technology (OSI network management) with another technology (CORBA) which is obviously not superior in this domain, is hard to justify. It is much better to position CORBA in an area in which it can obviously add tremendous value.

However, the complexity existed in the network and element modeling cannot be justifiably carried to the higher level of integrated cross-domain network and service management. The object simplification has to be made. In the cross-domain network and service management arena, there is no need to include all network details to the object model. The lower level details should all be encapsulated to produce a simpler and more abstract object model. In the later chapters, we specifically discuss these issues and assert that this simplification and abstraction are the key tasks to achieve integrated TMN management.

To address the second question we asked before, how to utilize CORBA technology in TMN, we examine more closely the application development issues by looking at the ways CORBA and OSI can work together to support TMN management functions.

6.5.1 Integrating CORBA and TMN

In preceding analysis, we have demonstrated that CORBA is an appropriate technology to support distributed network management functionality. Furthermore, CORBA is the strategic technology framework to support service management environments.

In the management of OSI networks, CMIP has distinct advantages to be used to manage large numbers of relatively unsophisticated objects representing the physical network. The best approach is to combine CORBA and OSI network management to provide an integrated solution so that the benefits of both technologies can be exploited.

The major problem with integrating CORBA and OSI-based TMN network management is the object modeling. These different technologies use very different approaches to model the managed objects. In OSI network management, a set of management object interfaces are defined using GDMO; the data types used in objects are defined in ASN.1. There is also a set of systems management functions defined in OSI standards. On the other hand, CORBA has its own object modeling technique. Its objects are defined in CORBA IDL (Interface Definition Language). A set of object services is also defined, by the Common Object Services Specification (COSS).

To integrate CORBA and TMN is to integrate these two different object models and the related management functions and object operations.

6.5.2 Modeling OSI Network Management Architecture

One of the issues relating to the CORBA/OSI integration is the reconciliation of distributed object architecture (CORBA) and the OSI network management architecture. One requirement targeted by some parts of the industry is that the integration must preserve the OSI network management architecture. To satisfy this requirement, the integration technology must create a complete GDMO object model in CORBA, such that all the following entities are mapped to CORBA space:

- GDMO objects and attributes
- Operations
- Notifications
- Containment topology
- Management entities (MEs)
- Communication functions

No matter which integration approach is used, we strongly resist the notion of creating a complete OSI network management architecture within CORBA through the integration technology. We believe that to use a truly distributed environment to represent a nondistributed peer-to-peer communication architecture is a backwards step.

To highlight this view, let's specifically look at the following OSI concepts:

- Containment structure
- CMIP filtering and scoping
- Modeling of application entities

6.5.2.1 Containment Structure. In OSI network management, GDMO objects model networks and elements. The number of relationships an object can participate in is limited, and the topology of objects is rather static. Based on this characteristic, the most important relationship among objects, containment relationship, is built as part of the object property within the GDMO model by using name binding facilities. The distinguished names are used as part of the object ID and also as part of the navigation device. The construction of the object containment tree is thus more efficient than that in the CORBA environment.

On the other hand, there is a fundamental difference in the concept of *relationships* between the CORBA object model and GDMO object model. The CORBA object model does not treat the relationships an object can participate as part of the object definition. Instead, it stores relationships in a separate service—the relationship service. The benefit of this approach is true encapsulation and reuse. An object should not have knowledge about which relationships it is in. The object can thus be free to participate in any relationships while preserving its core semantics. This concept is very important to the reuse of objects as building blocks for other services. However, the disadvantage of this flexibility is that the construction of a topology based on a certain relationship can be inefficient, as any such topology can only be constructed dynamically through reasoning with the relationship service.

Because of this characteristic, the containment tree is not a valid concept in CORBA. More importantly, an object ID in CORBA does not include any navigation information. This allows location independence to be supported in the CORBA object model.

This adds a problem to the integration. When using CORBA objects to represent all GDMO objects, the following approaches can be used to model the containment relationship:

- Use the CORBA relationship service to model containment. This is the more *natural* CORBA approach, but it is likely to be inefficient with large quantities of objects and their relationships.
- Build a containment tree using the OSI approach, i.e., to model the OSI concept of distinguished names. In this case, whenever an object is created (or deleted), the link to its superior object is created (or removed) by updating its superior object instances. This approach is a very crude way of using the CORBA object model.
- The third approach is to use the CORBA naming service. The naming service defines a concept of a naming context for an object reference, which may be used to model containment structures. However, keeping large numbers of object identifiers and naming contexts in memory is likely to be a problem.

No matter which approach is used, the containment structure is a very specific concept of OSI network management. The essence of this concept is that an OSI object instance is identified by its distinguished name and this distinguished name uniquely identifies the location of the object instance. This concept of combining an identifier of an object and its location is fundamentally different from the CORBA object model. Using this approach, a CORBA client simply cannot invoke an object request without knowing the location of the object.

This issue is a major concern of representing OSI architecture in CORBA. We recommend that no matter which approach is used to integrate CORBA and OSI, the OSI architecture should not be mirrored in CORBA space.

6.5.2.2 Usage of CMIP Filtering and Scoping. CMIP scope and filter facilities allow CMIP requests to be sent to objects based on their containment relationship and some other semantic conditions. A scoped operation first identifies a portion of the object containment tree as the search space. It then applies a filter to the object space and the filter allows the established containment tree to be traversed to identify objects matching the filter criteria.

This is a derived problem from the issue of containment tree discussed above. In the case where CORBA objects are used to model GDMO objects, it would be possible to represent the containment relationship between the objects using the relationship service. Scoped requests could then be sent to objects based on queries to the relationship service to establish the searching space, then the filters can be applied to each object found within the searching space.

While semantically this would provide for scoped requests, the performance of a system based on this approach is likely to be poor due to the large number of records the relationship service has to maintain and search in order to establish the searching space. On the other hand, this function is generically part of GDMO object modeling and is supported by the OSI architecture with reasonable efficiency.

Instead, if CORBA is not used to mirror OSI functions, a CORBA object can be used to model a portion of the network with a self-contained semantics and a set of related functions. The object modeling process can then avoid using scoped and filtered operations in the *interface definitions* of CORBA objects. Obviously, these operations can still be used in CMIP which forms the *implementation* of CORBA objects.

For example, a CORBA connection management object could represent an ATM network with a set of ATM switches and the physical connections. It has an operation `SetConnection()` as its interface. The CORBA object modeling can be done such that a scoped and filtered operation is not required on the CORBA connection object itself. Instead, in the implementation of this `SetConnection()` operation, it will require scope and filter operations on the switch objects and physical objects representing the network and these objects remain as GDMO objects.

6.5.2.3 Modeling Management Entities. The OSI management entities (MEs) are the basic software components used to construct the network management applications. In OSI architecture, an ME represents an end point of a peer-to-peer communication. Distributed applications can be built using these entities as components.

It is very plausible to model OSI management entities as CORBA object building blocks and use them as the units of distribution, encapsulation, replication, backup, etc. The integration technology can be used to model the boundary between these entities rather than the internal structure of these entities. Figure 6.5 depicts the OSI management entity interaction.

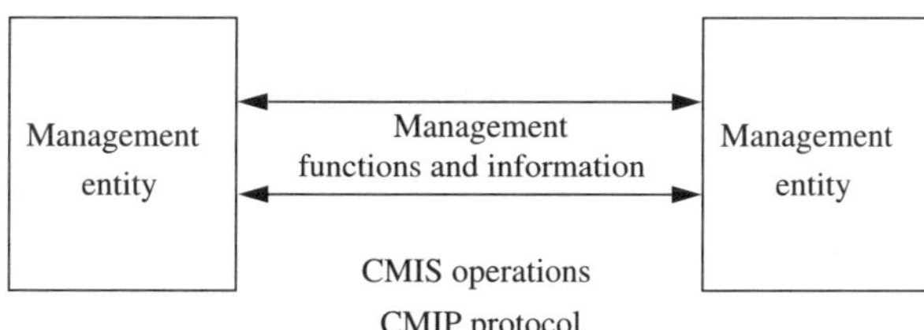

Figure 6.5 OSI management entity interaction.

Using the integration technology, this structure can be modeled as that depicted in Figure 6.6.

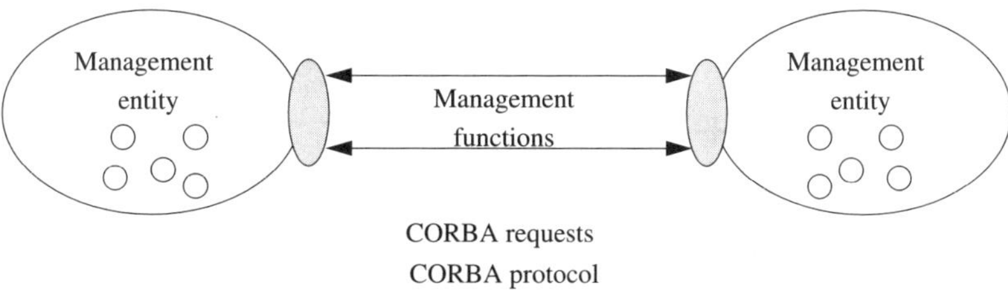

Figure 6.6 OSI management entity interaction in CORBA.

In this model depicted by Figure 6.6, a management entity in OSI is represented by a CORBA integration wrapper (shaded ellipses). These wrappers encapsulate the OSI management entities and the GDMO managed objects within them. The functions offered by the MEs are represented by the wrappers, and these functions and their interfaces don't have to have the same detail as the GDMO objects. They can represent more abstract functions offered by the ME. These interfaces manifest the network management functions (semantics) one can invoke on the ME, rather than only provide the peer-to-peer communication for exchanging object states. These objects can be modeled in CORBA properly; they don't have to preserve the OSI features such as containment, scope, filter, etc.

The model has changed the peer-to-peer communication paradigm to object request paradigm. The distributed CORBA-based application can be built using this interaction.

From the discussions in this section, we conclude that due to the following reasons:

- difference in object model,
- difference in application development, and
- difference in management structure,

CORBA may not be the right technology to replace the OSI in the element and part of the network management functions. OSI technology was specifically designed to address business and technology requirements in those areas; as such, it offers value that CORBA would find very hard to compete. Although some has demonstrated how CMIP/GDMO concepts can be implemented by introducing more CMIP-like features into CORBA, it is not convincing that the approach adds more value. In particular, the technical vision of the approach tangents away from the overall CORBA vision.

To summarize, we believe that *it is undesirable to model OSI CMIP architecture in CORBA even though it can be done.* Rather, it is more plausible to use CMIP communication facilities and management functions to manage the low-level network activities and use CORBA to represent more abstract functions based on the underlying CMIP functions.

On the other hand, in the integrated cross-domain network management and particularly the service management area, CORBA finds its value by adding more flexible distribution and scalability, and ultimately it can act as a technology integration framework to bridge the gap between the management of telecom resources and the management of telecom business and customers. In addition, it can reduce the intrinsic complexity of the object modeling in the network and element management by supporting the simplification and abstraction in the service and customer management areas.

All these point to the requirement for a sound integration strategy between CORBA and OSI technology. In the following chapters, we discuss in more detail the business requirements for technology integration and present the different integration strategies and object modeling approaches to achieve the integration.

6.6 INTEGRATED NETWORK MANAGEMENT—A CASE STUDY

In this section, we provide a simple example of building distributed TMN network management systems that support both CMIP and CORBA management interfaces.

This example is used as a prototype [27] to demonstrate that CORBA technology can be used to build large-scale distributed TMN network management applications and it can be used as an integration framework with the existing OSI management applications.

The prototype deals with the management issues in a large network management application which has potential to scale and distribute in a large geography and across multiple technology domains. It supports an ITU-T G.805-based network hierarchy [86] that is scalable and distributed. It establishes the concept of generic network model to provide an abstract view for supporting network technology independent management applications. It is a CORBA-based system yet supports the integration to CMIP element managers to demonstrate the integration concepts discussed before.

6.6.1 Overview

The purpose of this research is to formulate the distributed architecture for a network management system (NMS) capable of managing a multidomain, multivendor communication network. That is, the NMS would be capable of managing a physical network consisting of heterogeneous transmission equipment (ATM, SDH), where that equipment, even within the same domain, could be produced by different vendors. Figure 6.7 depicts the high-level concept of the NMS.

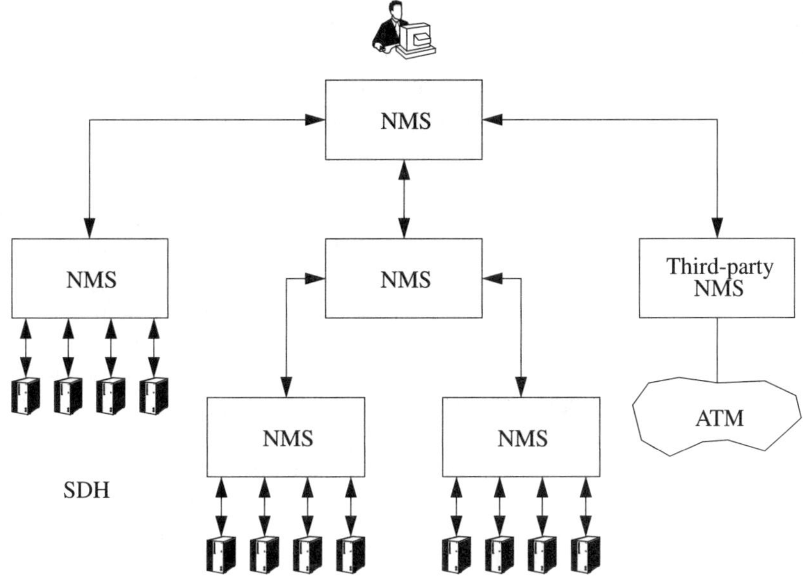

Figure 6.7 NMS for management of a heterogeneous network.

The functionality of the NMS includes:

- *A network representation*—using a standardized generic model
- *End-to-end connection management*—including automatic, cross-domain routing, and equipment configuration
- *Fault management*—faults reported from managed equipment would be translated into the generic network model and propagated upwards

The following design goals were applied in this research:

- *Support for recursiveness*—A hierarchical arrangement of management nodes was chosen to maximize the scalability and flexibility of the deployed system.
- *Support for reusability*—It was determined that the same software component should be present at each level of the hierarchy, rather than differentiating between leaf, branch, or root nodes of a recursive network architecture. This would reduce development and maintenance costs by decreasing the number of individual software components.
- *OSI and CORBA interfaces*—Rather than restricting the north-facing (to the management component) interface of the management node to a single technology, the architecture would be flexible enough to support both OSI and CORBA management domains in parallel.

6.6.2 Network Model and Management-Facing Interface

The NMS manages networks from multiple domains such as ATM, SDH, Frame Relay, and ADSL. It is also desirable to present a single view of the entire managed network. A generic model was chosen to represent all of the network domains, based directly upon the functional architecture described in ITU-T Recommendation G.805.

This model uses a small set of conceptual objects to represent any transmission network. The model is split into layer networks, with each layer representing the capability of transmitting a fixed kind of characteristic information (for example, SDH STM-N signal, SDH VC4 virtual containers, ATM 53 octet cells) across that layer. Each layer contains subnetworks, which may be partitioned into smaller subnetworks. A subnetwork represents a collection of connection points, any of which may be cross-connected with any other connection point on that subnetwork. A leaf subnetwork can represent a piece of transmission equipment (such as an ATM switch), or a subordinate management domain.

For the OSI management domain, a north-facing (management facing) CMIP interface is provided, using the GDMO/ASN.1 MIB I-ETS 300 653 [73]. This MIB provides an implementation of the abstract modeling concepts described in ITU-T Recommendation G.805. While it is suitable for modeling an SDH network without change, some minor modifications were necessary to allow an ATM network to also be modeled. The MIB provides M-Actions for setting up and releasing trails (end-to-end connections across a layer network) and subnetwork connections (a cross-connection across a given subnetwork within a layer network). Changes to the contents of the instantiated MIB, as well as fault events (such as communication alarms) are emitted as M-Event-Reports.

A CORBA IDL interface is designed, which is based upon the modified version of the I-ETS 300 653 MIB. The interface specification loosely corresponds to the original GDMO specification with a degree of abstraction. Figure 6.8 depicts the management interfaces supported by the NMS.

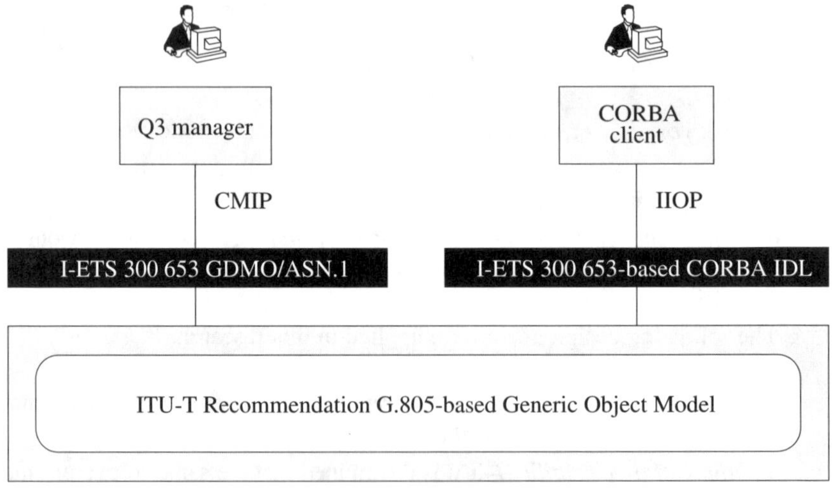

Figure 6.8 The north-facing interfaces provided by the NMS.

6.6.3 NMS Systems Architecture

The overall NMS systems architecture is depicted in the Figure 6.9. This NMS architecture consists of the following components:

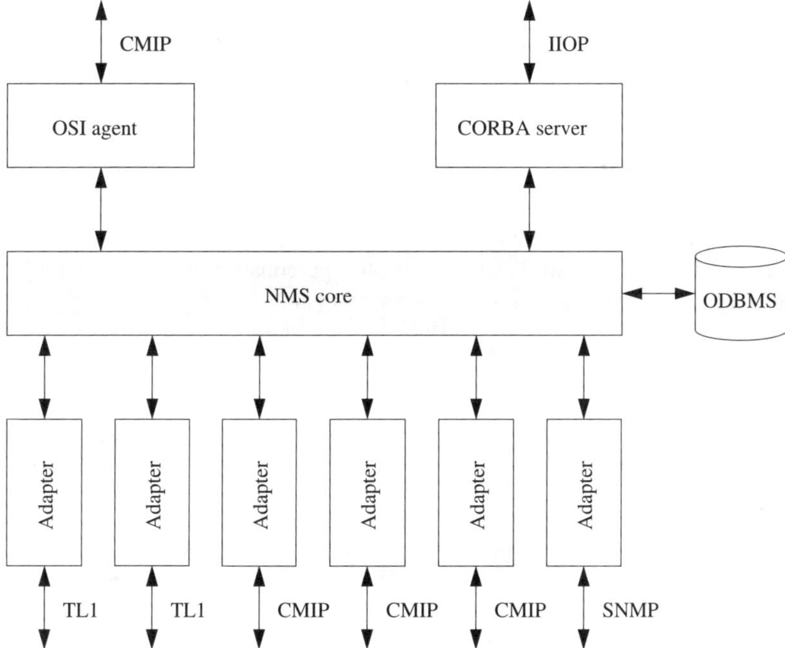

Figure 6.9 The NMS architecture.

- *NMS core*—This component maintains the generic network model. It offers a domain-independent interface to the OSI agent and CORBA server components. The contents of the network model are made persistent using an object-oriented database. The NMS core also contains the functionality necessary to perform path management, specifically a routing mechanism for establishing a connection across multiple domains. The core remains abstract, not possessing knowledge of specific vendors' equipment or management systems. Instead, it uses the small set of domain-specific interfaces offered by the *adapter* components.
- *OSI agent*—The OSI agent offers the north-facing CMIP interface, based on the specialized I-ETS 300 653 MIB. The containment tree it maintains is derived directly from the domain-independent interface offered by the NMS core. Any changes made to the containment tree are translated into modifications in the generic network model through the same interface. Events received from the NMS core are likewise translated into m-event-reports.
- *CORBA Server*—The CORBA server offers the north-facing IDL interface derived from I-ETS 300 653. As with the OSI agent component, the CORBA objects maintained by this component are based directly upon the contents of the generic network model in the NMS core. Operations on the CORBA objects are passed on to the NMS core. Events received are emitted through an OMG COSS Event Service event channel.
- *Adapters*—A minimal, generic set of interfaces are used to represent the various entities that can be managed, which include SDH and ATM equipment, domain-specific management systems, as well as subordinate NMS systems. To communicate with actual instances of these entities, an adapter component is implemented. This presents the generic, north-facing interface, which allows it to be integrated with the NMS core. The adapter implements this interface by communicating using protocol

that is appropriate for the entity being managed (this could be TL1, SNMP, CMIP, or CORBA). The NMS can thus be extended to support new equipment or underlying management systems by implementing new adapters, and associating them with the NMS core.

6.6.4 Implementation

A realization of this architecture has been implemented in a prototype, whose architecture is illustrated in Figure 6.10. This prototype consists of the following software components, each of which has been implemented in C++:

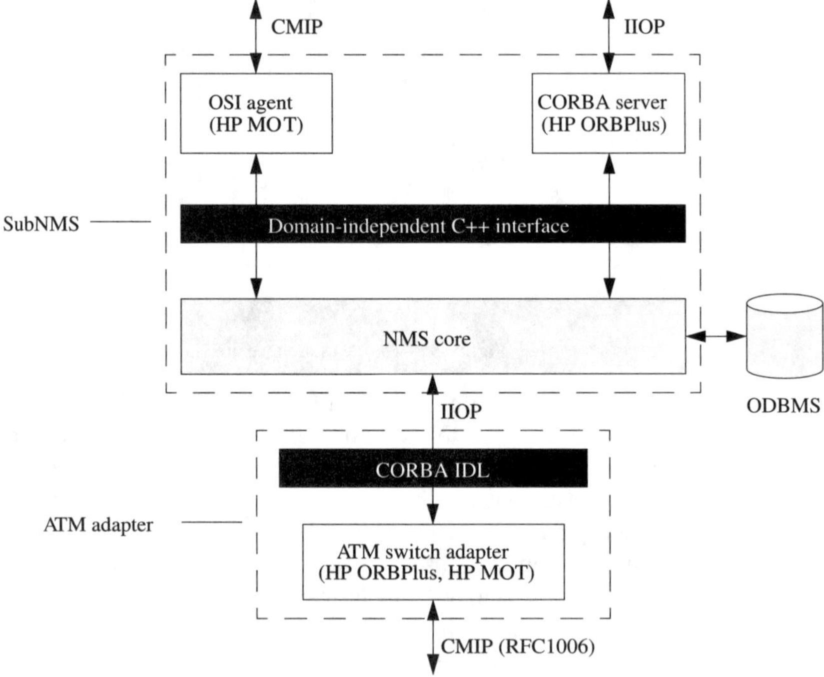

Figure 6.10 The NMS implementation.

- *SubNMS*—The SubNMS incorporates the NMS core, OSI agent, and CORBA server components. It runs as a single UNIX process. The domain-independent interface to the NMS core was modeled using C++ objects. Both the OSI agent and CORBA server components integrate with the NMS core through this interface. The network model maintained by the NMS core is persistent by using an ODBMS.

  The OSI agent component is implemented using HP's MOT (Managed Object Toolkit). This tool allows for the rapid development of OSI agents, and proved to be very effective in constructing this component. The OSI agent runs in a single thread within the SubNMS process.

  The CORBA server component uses HP's ORBPlus, a CORBA2-compliant ORB implementation. The facilities offered by this ORB, especially the object activation mechanism, allowing for the on-demand instantiation of object implementations, allows this component to be developed quickly and effectively.
- *Adapters*—Each of the minimal sets of adapter interfaces were designed using CORBA IDL. Each adapter was implemented as a stand-alone CORBA server. All communication with the SubNMS occurs via the ORB.

For the prototype, an adapter was constructed for the management of an ATM switch. An OSI agent, implemented using the OSIMIS/ISODE CMIP stack [54], was developed to manage the ATM switch. The adapter was implemented using HP's MOT to generate the OSI manager-side code necessary to communicate with the OS-IMIS/ISODE agent. HP's ORBPlus was used for the north-facing CORBA interface to the adapter.

6.6.5 Deployment

The deployment used to demonstrate the viability of the NMS architecture consists of a two-level management hierarchy, managing an ATM network.

The ATM network elements are divided into two four-node subnetworks. Each subnetwork is managed by separate SubNMS processes, which are run on separate hosts.

An ATM switch adapter is co-located with each SubNMS. The adapter is used to communicate with the ATM agents. There is one such agent for each ATM network element, running directly on the ATM hardware. The ATM network elements from each subnetwork are connected to their management station via TCP/IP running over an ATM virtual path, which is preconfigured for management communication. The ATM switch adapter communicates with the agent using RFC1006 (CMIP over TCP/IP).

Each ATM subnetwork is managed directly by interacting with the SubNMS responsible for that subnetwork, through either its OSI agent or CORBA server interfaces.

A third SubNMS is then installed on another management station. It is configured to manage the two subnetworks modeled by the existing SubNMS processes. This management station is connected to the other SubNMS stations via TCP/IP running over ethernet: the top-level SubNMS interacted with the subordinate SubNMS systems through the subordinates' north-facing CORBA server interfaces.

The view presented by this top-level SubNMS, through its own north-facing OSI and CORBA interfaces, is that of two ATM network domains, with no details visible. That is, the individual ATM network elements cannot be discerned through this interface. The view exported from the subordinate SubNMS systems to the top-level SubNMS consists of an ATM *cloud,* only the external, connectable access points are visible.

A simple Java GUI application is implemented to interact with each SubNMS. The Java GUI communicates with the SubNMS systems' north-facing CORBA server interfaces, using IIOP running over TCP/IP. The Java GUI is associated with the top-level SubNMS to allow the high-level domain view. It can also be associated with the subordinate SubNMS systems to display their more detailed element management view.

6.6.6 End-to-End Path Setup

The primary functionality of the NMS is to perform end-to-end path management operations. Figure 6.11 illustrates the path creation functions of the NMS.

To create a connection in the deployed system, a request is made through the top-level SubNMS to establish a virtual path (VP) between two access points, where each access point is associated with an ATM network element from both subnetworks being managed.

The request is directed to the north-facing CORBA server interface of the top-level SubNMS. In the CORBA server component, it is translated into a request onto the domain-independent C++ interface offered by the NMS core component. The routing function is then activated. This determines that a route can be established across the two subnetworks that represent the managed network domains controlled by the subordinate SubNMS systems. The request to establish the individual connections across both domains is then dispatched to the

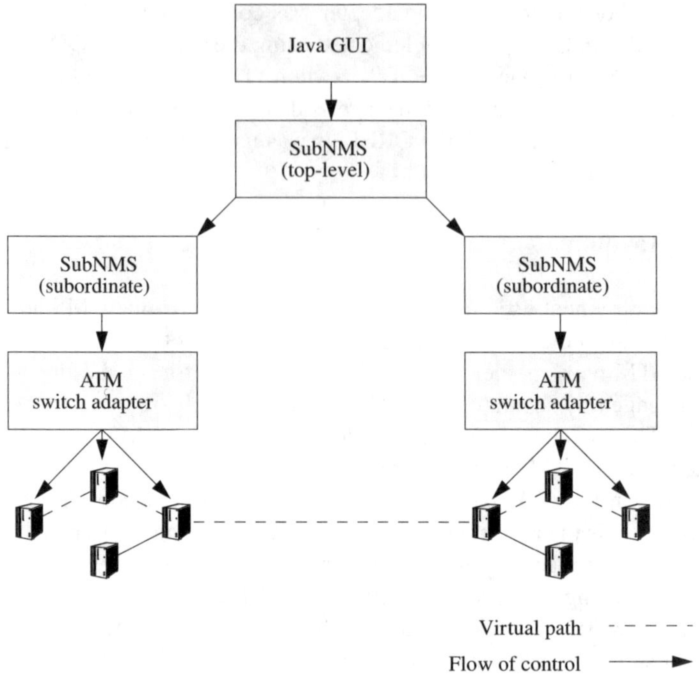

Figure 6.11 Setting up a VP using the prototype NMS.

subordinate SubNMS systems in parallel. The thread performing the path setup then blocks waiting for both requests to complete.

The request is received by each subordinate SubNMS simultaneously and is processed in parallel on both management stations. Each SubNMS activates its own routing function and finds a route between the supplied access points within their own network domain. This translates into setting up subnetwork connections across the subnetworks representing the ATM network elements. Once the routes are found, each SubNMS then dispatches the requests to establish the required cross-connections to their associated ATM switch adapter in parallel.

The ATM switch adapters translates the cross-connection requests into a series of m-creates and m-actions which are directed to the agents controlling each ATM network element. The appropriate configuration changes are made to the ATM hardware. The successful operations are reported back to the adapters, which in turn propagate results along the nested subnetwork structure back to the top-level SubNMS. The original caller is notified that the VP is successfully created.

Once each SubNMS had successfully established the required connections in the underlying domain or adapters, the network model each maintained is updated to include an appropriate representation of the path or cross-connections that has been created. Details of all existing and newly created paths are visible through either the north-facing OSI agent or CORBA server interfaces.

6.6.7 Event Propagation

Where the end-to-end VP establishment shows that top-down management can be accomplished, it is also necessary to demonstrate that bottom-up information flow can also be effectively propagated upwards.

After establishing a number of VPs through the top-level SubNMS, a fault is introduced into the system to represent disconnection of the fiber cable between two of the ATM network elements.

The failure of the physical connection is detected by the ATM network elements at both ends of the cable. The alarms are passed onto their respective agents, which emit CMIP `m-event-reports` indicating the problem.

These `m-event-reports` are received by both ATM switch adapters and are mapped into a CORBA event and passed through the adapters' IDL interfaces upwards to both subordinate SubNMS systems.

When the subordinate SubNMS systems receive the events, they are applied to the network model that each SubNMS maintains. This results in a number of objects contained within the generic network model to change state. These include those objects representing connections affected by the faulted physical connection. This causes events to be emitted from the domain-independent interface supported by the NMS core. The OSI agent and CORBA server components then catch these events, which are mapped into `m-event-reports` and OMG COSS Event Service events, respectively.

The CORBA events, transmitted through an event channel (one event channel being used per SubNMS) are received by the top-level SubNMS. These changes are applied to its own generic network model, which likewise causes events to be pushed upwards and emitted through its own OSI agent and CORBA server. In this way, the objects representing the VPs passing over the faulted physical connection in the top-level SubNMS become notified. The fault conditions are displayed in the Java GUI associated with the top-level SubNMS.

6.7 SUMMARY

In this chapter, we demonstrated that CORBA, as a general purpose distributed computing technology, can also be used as distributed management technology for different TMN management tasks.

We demonstrated that, although it is achievable, it may not be desirable to completely replace OSI element management by using CORBA. CORBA can easily contribute to TMN management in its own way by adding value in different areas. The integrated end-to-end network and service management is the primary target area where CORBA can add considerable value.

The case study has demonstrated the suitability of using CORBA to implement distributed network management functions and to integrate these functions with the existing OSI-based element and network management systems. In this example, CORBA offers an abstract view for network management information and functions. This view is important for the purpose of an end-to-end focused management.

The study also suggests the importance of the generic object modeling for network functions in order to achieve the cross-domain management functions using CORBA. In the following chapters, we will demonstrate that this object modeling is not only the key to achieving the seamless integration between different TMN layers, but also the key to achieving the technology integration between CORBA and OSI.

7

Integration Requirements

In the previous chapters, we discussed the need for business process reengineering in the telco industry by introducing more process automation and information flowthrough. This can be achieved by using more established IT technologies such as distributed OO and Internet to implement TMN architectures and applications. We also discussed the requirements, principles, and some examples of building large distributed TMN systems using distributed OO technology such as OSI and CORBA.

However, one fundamental question facing the telco software industry is, *What IT technology should we use to create solutions?* This question leads us to have a close look at the requirements for the IT technologies and their integrations [21].

7.1 TECHNOLOGY WARS

The last decade has seen the proliferation of mainstream IT technology in the telecommunication industry as the consequences of the business process reengineering. This has increased the level of automation of the management software in an industry which has created a fully automated global telephone network. The IT technology has strengthened the telco's operation support systems and enabled the telephone and data networks to be more efficiently managed.

The last decade has also seen the global proliferation of international standards effort. This effort stems from the needs to build real *global* systems: voice and data networks and management systems. The *open systems* movement as represented by the ITU-T and ISO, in particular the *Open Systems Interconnection* (*OSI*) initiative, has largely contributed to this industrial trend to build networks and management systems which are based on open standards so that they can be constructed in a *plug and play* fashion. Obviously, the perceived economic benefits are that this will drive the system cost down as it greatly encourages the vendors to produce better and standard conformant *plugs* which form parts of the total solutions. This will ultimately present the telco industry with procurement choices from different vendors for their network equipment and software components.

By and large the theory worked (to a degree). The international standards (de jure and de facto) for both telecommunication networks and IT software architecture and technologies for management flourished. Software architectures for managing telecommunication networks have been produced and products to realize these architectures have been made available to the industry.

The technology development in the mainstream IT industry played an important role in pushing the management technology boundaries of the telco industry. We have seen that all aspects of IT have made their way into the telecommunication industry and have influenced the industry in various degrees. These include the operating systems, computer languages, object-oriented specification and design methodologies, protocol specifications, formal specification and description techniques, the distributed computing architecture, workflow management, the GUI technology and finally the Internet, Web, CTI, IVR, etc. All of a sudden, we realized that the telecommunication industry, although having some unique characteristics, by and large has no fundamental differences from other industries where IT technologies have been successfully used in the past.

However, the benefits of the modern IT standards and technology also come with the cost and confusion. First, the IT industry itself is moving forward at the frightening speed, with its emergence of new technologies and products. All these new technologies come with different functional manifestation and different levels of maturity. This has created huge difficulty for the telco industry to find suitable IT technologies to manage its main business: provide telecommunication services to the customers. This has led to the first question facing every IT manager in the industry:

- how to choose the IT technology to support the business solutions and the business process reengineering—which is the winner?

A typical scenario is that the IT manager responsible for the business process reengineering hires an IT technology consultant to help selecting technologies for the business process reengineering. The IT industry was very hot on three-tiered client–server technology at that specific time. The consultant managed to convince the IT manager that the three-tiered client–server technology was the right choice. The IT manager was convinced and spent the next six months (this is the lightening speed in the industry) to convince their peers and superiors. Finally, when all the support was obtained, checks signed and ready to go, the consultant came back telling them that the three-tiered client–server technology is out of date, it is now the Internet/Intranet technology!

Furthermore, because of the technology evolution, telco is faced with the problem of integrating different operation support systems. These systems may have been developed at different times for achieving different functions and using different technology, or they come through merges and acquisitions. To integrate all these systems is a nontrivial task for any operators.

The confusion in the standards arena is no less severe. The emergence of multiple global and regional, general, and specialized standards bodies have also created huge difficulty for IT managers in the telco industry. The first question many IT managers ask is, Which standard specifications should be followed? The reality is that in almost every aspect of the telecommunication industry, there are over supply of standards and specifications. There are multiple standard definitions of transport architectures; there are multiple standard object definitions for the transport technology, some are specific and some are generic; and there are multiple protocol definitions for exchanging information, etc. It is much worse in the software industry where there are always multiple standard ways of doing the same thing, including the overall frameworks of distributed computing, and the protocol details of information passed between different sub-systems in a business transaction.

All these have created and contributed to the so-called *software technology wars.* There are wars in almost every area of the telecommunication management industry. There are classic wars of CMIP and SNMP in the management protocol area. A few years ago, every major international conference in network management (such as IM and NOMS) witnessed wars between CMIP and SNMP. More recently, there are wars in the management middleware: DCE, CORBA, OSI, JAVA, DCOM. There are wars in different object modeling methodology and tools: GDMO, IDL, ODL, OMT, UML, DMI, etc. There are wars in different distributed application architecture: OMG's CORBA, OSF's DCE, TINA-C's DPE, Microsoft's DCOM, Sun's EJB, etc.

In the area where war is not so open and nasty, there are also many different specifications of similar functions. There are different but similar object specifications for the connection model in ATM Forum's M4 specification and ETSI's I-ETS 300 653; there are different but similar service management functions and architecture between ITU-T, TINA-C, and TeleManagement Forum; etc.

Technology wars are caused by rough and tumble free market forces. They often lead to healthy competition and better products. But sometimes all they lead to are delays in solving problems. Technology wars can be good, but only when we manage their impact. Unfortunately, in many cases, people don't know how to manage the impact. People are dragged into these technology wars and forget why we are here in the first place.

7.2 AVOID THE IMPACT OF TECHNOLOGY WARS

Let's consider an issue bigger than the technology wars, namely the life-cycle management of software development for the telecommunications industry. Since this industry has been at the leading edge of the adoption of software technology, its experiences are often carried directly into other industries. To manage the software development life cycle, one must manage the life cycles of its paradigms and products.

A paradigm is a way of solving problems. It is a belief system that the software activities should be conducted in a particular way. The typical examples of paradigms include batch processing, mainframe architecture, relational database, artificial intelligence, and more recently and more relevant to this discussion the distributed object computing. Paradigms typically have a much longer life cycle. They typically have three stages.

A paradigm almost always spends its infancy in universities and research institutes. During this period, there are lots of wars but largely irrelevant to the industry. Because it is far too early to change any of industry's software development behavior before this stage is complete. Many paradigms die before reaching this point.

When a paradigm reaches to a mature age, people start to build products and software solutions. This is where war starts. Some technology or products will emerge as the winners of the war and some will not. This is where standard processes start which lead to more wars. When we reach the apex of the maturity, we start to see the *dummy's guide to XYZ* in the bookstores.

In time paradigms die. Batch processing, once a most powerful computing paradigm has died gloriously. Today's paradigm is tomorrow's legacy. There is nothing wrong with this. We just have to manage this process.

Paradigms are realized by technologies and products. At any stage of the life cycle of a paradigm, there are different technologies and products with various levels of sophistication, maturity, and customer acceptance. Figure 7.1 depicts the relationship between paradigms and technologies.

To avoid a technology war, we need first to recognize that we are building a business solution that will outlive individual technologies and products by a large margin. We need then to consider how to minimize the negative impact of the technologies and products on the longevity of our solutions. Further, we need consider whether we can spend all our effort to pick the best technology and then adapt our solutions to it. The challenge is to find the technology that will naturally evolve in step with the needs of the business.

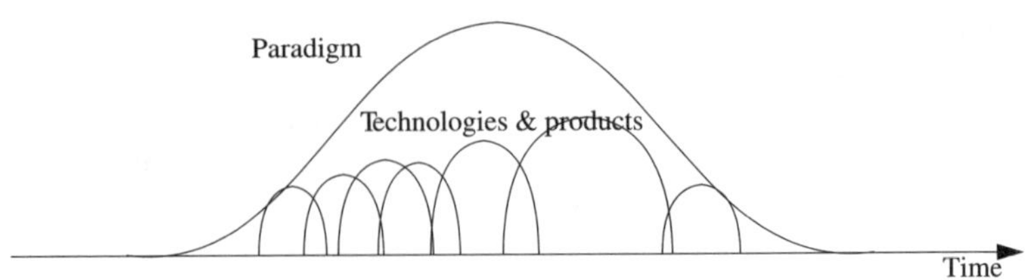

Figure 7.1 Life cycle of paradigms and technologies.

A paradigm may be flexible enough to encapsulate some features and concepts of an older paradigm, a product is never meant to bridge two paradigms.

When people build today's business solutions, they must be responsible for today's operation and for yesterday's systems. This holds for even those people building for new telecommunications companies born from global deregulation. Even they have a green field opportunity, they must fit into the existing worldwide telecommunications infrastructure. Today is tomorrow's yesterday. There is no shortage of examples of some new telephone companies having to face the past with the same level of difficulty as the long-lived ones.

Although our major responsibility of building software is for today and yesterday, it is obvious that people who don't consider tomorrow in their solutions are living an irresponsible life. However, we go further to state that people who think they can build solutions for tomorrow (next five years) are also living in an unreal world, thus also being irresponsible. We cannot impose today's paradigm to the builders of the future solutions because paradigms do change and tools supporting the paradigms change with them. Using today's language to implement yesterday's paradigm is not the right way to manage development life cycle. Our best option is to state exactly what we are building today, a clear interface specification of our system and components separated from the actual implementation and algorithms. This is our best chance that tomorrow's software builders will understand what we have built and try to integrate our components into their solutions.

7.3 COMMERCIAL APPROACHES TO STANDARDIZATION

One of the major battlefields for technology wars is the standards process. So far, there are two styles of standardization, the totalitarian and democratic. No matter what the styles are, two elements must exist to have success: the *moral* authority (people trust that the technically best is produced); and the *commercial* authority (that the commercialization of the standards can be enforced). So far some standards processes cannot achieve even one of these. As a consequence, the processes that all set out to reduce the technology wars intensify them.

How many times we come across an RFI or RFP which states that the solution must conform to a large list of open standards. Can those who write the RFI/RFP elaborate convincingly that there is a real business need to conform to these standards and they really believe that their needs can be met? Or it is just part of being a good *corporate citizen.*

Some modern standards processes are flawed in at least two areas (apart from the lack of moral authority and commercial authority). First, we try to standardize things before a full free market competition is well underway and people's creativity and innovation are exhausted. At this stage the value propositions of standards processes and the commercial world are different. The commercial world aims to produce new products with feature and function differentiation, and the standards processes expect the commercial world to agree on ways to make their product commodities. The result and the reality are that many standards efforts become international collaborative research so they lose their moral authority. Second, we set a goal of plug and play (and we sometimes mislead the users that the plug and play can be achieved for free) or full integration. Plug and play can only be achieved when a thing runs out of its commercial value and becomes a true commodity. Whenever someone has the goal of producing the best X, they need to break out of the current paradigm. This often cannot be done in a plug and play manner. Furthermore, it has been proven, if not done properly, the extra cost of plugging different components for solutions can be huge.

So how to avoid wars in the standards world? Are open standards useful? Yes, but as an architect responsible for delivering business solutions, the conformance to open standards should never be used as the criteria for success. To build a solution that minimizes the impact of technology wars is our success criteria. In achieving this goal, we may choose to conform to those open standards that fit the problem.

7.4 MANAGING BUSINESS SOLUTION LIFE CYCLE

The key issue of technology integration is the management of the solution life cycle. As discussed before, we need to separate the different life cycles and determine which life cycle should coincide with the telco industry's business process reengineering.

7.4.1 Application Integration and Systems Interoperation

The open standards movement in the past decades aims to achieve the systems level interoperability. This interoperability is manifested at the protocol level, i.e., two component systems exchange information in a predefined and commonly understood format. The benefit of this interoperability is that different components can be manufactured by different vendors so long as the common protocols are faithfully implemented at each end of the information pipe.

So far this has largely impacted the telecommunication physical networks. If the common object model and transport protocol are used, different elements on the different parts of the network are able to understand the information passed between them. As such, different network equipment can be interchanged (this is still a half-realized dream) without affecting the business activities they support.

However, even if this level of interoperability is completely achieved, there is another bigger issue which remains completely unsolved: the *application integration.* The components of the management applications don't talk to each other. Each of these components is responsible for a portion of total management functions and services, but they remain isolated and not interoperable with each other.

Larry Bernstein in his keynote presentation: *Can we talk?* at ISINM '95 in Santa Barbara has illustrated this phenomenon. The global telephone systems have achieved complete interoperability. One can pick up the phone from the United States and call a number in Japan. Within a few seconds, the systems are interconnected and we can exchange signals. However, we still cannot talk since two speakers may not understand each other's language!

This problem is typically characterized by the *change of address* request to the service provider. When we moved from our house a year ago, it took us three times to update the address records with our local service provider for a very simple service portfolio. Just when we thought we had notified the necessary authorities, we still received letters sent to our old address from the section handling credit card which was linked to our telecard! When we explained that we had already notified the billing department and telecard section, we were told that this is yet another separate department! (Has this got anything to do with us?)

The application integration is a key to the notion of becoming a smooth operator, and this should be the first goal of the business process reengineering. It is also the key to the service differentiation to win more customers by achieving better customer management. How many customers care about five TMN management functions? All that customers want is to be able to order services, get them delivered, have some control over the quality of service management (report troubles, know the QoS status, get notified for likely problems and remedies), and receive an easy to understand bill with no surprises. More importantly, they demand an integrated management for all their accounts with the service provider. They don't have to receive reports from each of the departments representing different applications and services.

The application integration requires an integrated business objects and processes modeling. It is hard, but it is more important than element level interoperability.

7.4.2 Technology Management

In the world of technologies explosion, every software practitioner with responsibility of building business software should ask the following questions:

- What is our fundamental strategy towards achieving the technology coexistence?
- What are the successful criteria?

First, let's look at how a software project is managed under today's software engineering practice. At the very beginning, there is a business problem: we need to offer a new service and we need it quickly. We allocate a budget, put together an RFI/RFP to kick off the bidding process, selection process, and finally sign the contract. When we get to the software development, we spend one year developing a requirement analysis, we freeze the requirement, freeze the design, we then spend two years to build the software, test it, and put it to operation and it fulfills (or does not fulfill) our original business requirements.

After or sometimes before this process has completed, we would have other business problems emerging. We start another cycle of the process to deal with them.

This is the life cycle of a project or product. Who is managing the life cycle for business solutions in this exercise, nobody! Who is in control, the chief architect responsible for delivering the solution?

It is more likely that the control lives in our financial and accounting system, our contract system, our legal system; the whole control has only one purpose—that the software be delivered and nobody gets sued.

This is the control. The whole society's business environment, financial environment, and legal environment plus the software management regime (the IEEE, ISO9000, software quality, all those hitchhiker's guides to software development every project manager has to read from day one) control the outcome of the software development.

This segmented and discontinuous control model subjects us to the full negative impact of the technology wars. We all set out to build solutions using best technology and products; but end up building legacy systems which are stand-alone and nonintegrated.

The business solutions normally point to long-term objectives. Different parts of the objective at different stages are achieved by individual projects. In a well-managed engineering environment, individual projects can be well managed. However, there is no evidence and certainly no agreed upon software engineering practice to manage the whole life cycle of business solutions. The business goals are normally following the directions of a collection of individual projects using different technologies. This practice means that the progress toward the final business objectives are randomly made and there is no guarantee that all the individual efforts will converge to the final goal. Figure 7.2 illustrates this phenomenon.

The concept of technology tangent refers to the fact that the path to the final business objectives is not a straight line (as previously planned). This is due to the evolution of technology and individual circumstance of each project that contribute to the final goal. However, each individual project may tangent away to a different direction; the technology management role should be to make sure that they don't go too far from the business directions.

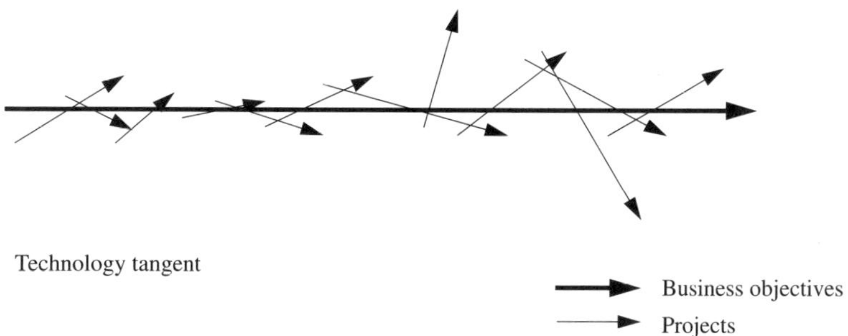

Figure 7.2 Technology tangent.

The responsibility of technology management is to manage this process. It needs to reconcile sometimes conflicting activities: the pursuit of new and better technologies in the business solution components and the achievement of the business objectives. When a project decides that CORBA is the best software architecture to implementing the service management systems which require a distributed platform support, the project team has a very legitimate reason to do this. However, someone should consider what if CORBA technology dies in three years, or what if Microsoft's DCOM wins the market (either by its technical superiority or by its sheer market size), or the whole world has moved to Java and Enterprise JavaBeans architecture. Someone should make sure when these events happen, we are still able to manage our business, services, and customers. Our business shall not be rendered unoperatable only because Microsoft's fortune has changed.

More importantly, technology management is the recognition of coexistence. Technology management is not the constant quest for using the best technology to implement the business process reengineering. The purpose of good technology management is indeed to avoid periodical and expensive BPRs!

7.5 RIGHT TECHNOLOGY FOR THE RIGHT DOMAIN

The telco industry has gradually recognized the multiple technology coexistence. In the existing telco OSS systems, different technologies are used to provide different functions. For instance, the proprietary protocol definitions might still be used to model the network elements; the element management might perhaps be provided by using a mixture of TL1, some proprietary mapping, or SNMP/CMIP; the network management systems might have been CMIP-based TMN systems, they might or might not interoperate with each other since they were provided by different vendors. The service management system might even be more a mixture of everything: from a number of billing records databases (some prerelational, non-SQL), to some Web-based customer ordering systems to interact with customers. Furthermore, many telco *functions* and *services* might be implemented as self-contained vertical systems—each had its own software architecture, middleware technology, GUI look and feel and its own database to keep the customer information, service records, and application data. In reality each of these services reflected the telco's organization structure: the mobile service section, the IN service section, the ISDN service section, etc.

The technology coexistence is also reflected in the fact that different technologies are used by different service providers. They did not have to work together much, but now this is all changing. With the new round of merges and acquisitions, many service providers end up with an even more mixed operation support systems base, all supported by different software architectures and technologies. Furthermore they are forced by the rules of deregulation (such as FCC interconnection regulations) to work together, which creates another technology coexistence problem.

The reality of technology coexistence has forced the telcos to rethink their entire management systems and OSS development and deployment strategy. It has forced them to abandon the notion of the *best management technology* for building OSS systems, and encourages them to consider the value and maturity of different management technologies and to focus on the integration of these technologies to achieve business solutions. This is the reason why *right technology for the right domain* is becoming a catch phrase in the telco industry.

There are different management problems in different telco support systems. The network elements need to be modeled and managed, the regional and global networks need to be managed at different levels and with different granularity. The service management functions such as service order, provisioning, quality of service, and billing need to be managed. Finally the customers require a management system where access, enquiry, order, service delivery, QoS tracking and billing enquiry and payment can all be supported in an easy to operate manner. Furthermore, all these subsystems need to be integrated as far as data flow and process flow is concerned to achieve the maximum automation and consistency.

In recent years, many discussions have occurred in the international conferences and forums relating to the issues of what is the right technology for what domain? At the center of the attention are three candidates, but they obviously represent a large range of alternatives:

- OSI as the incumbent technology for implementing TMN element and network management systems
- CORBA as the de facto architecture for implementing large-scale distributed system, which can be applied to support network and service management functions and OSS systems
- Java, Web, and Internet as the technology framework to manage customer services and access to OSS functions.

Other candidates include SNMP, Microsoft's DCOM, etc. Also there are many debates regarding which technology performs better in each domain, such as some in the CORBA camp advocate that CORBA is the platform to implement both network management and element management, as well as its well-accepted role in the service management area. Java has also been advocated to be the replacement for CORBA to support not only access to OSS but the component-based software architecture to develop and deploy OSS functions.

Perhaps the most important and difficult issue facing the industry is how to integrate these technologies to offer a solution. This is a difficult issue because any integration of multiple systems or technology requires some agreements on the underlying information model and process model. All technologies mentioned before are tools to build telco solutions; it is therefore meaningless to discuss integration without some understanding of the *contents,* which is the target of integration.

7.6 AN INTEGRATION EXERCISE

The TeleManagement Forum, in its April 1997 general conference at Long Beach, California, devoted some effort on this issue. TMF (then Network Management Forum) organized a Strategic Architecture Roundtable which brings all the senior technical and business representatives from its member companies. These companies represented a wide cross section of the industry including service providers, equipment vendors, and independent software vendors (ISV).

As a global telecommunication industrial forum aimed at achieving the smooth operation by focusing on the process flow from customer facing service and business management to equipment facing element management, TMF faced the challenge of implementing this process flow in the multiple software technology world. On the one hand, the members representing software suppliers have been continuously promoting the most advanced management architectures and solutions (sometimes not consistently). On the other hand, the service providers and equipment providers have been continuously requesting assistance from TMF to help them to utilize these technologies to provide solutions to their customers.

With this background, TMF aimed to achieve a strawman's proposal for a right partition of software technologies into the telco domain and the first attempt to define an integrated architecture. At the strategic roundtable, the participants were given a number of technology options from which they were asked to choose the most likely technology solutions for each layer in the TMN model and to justify why that option would be chosen by the industry as a whole over the next few years.

The options were:

- Mixed integrated technology solution
- Java-based solution
- CMIP-based solution
- Any other solution

The participants who chose solution 1 (75% of the group), were split into three separate groups and independently came up with the same breakdown of technologies for each layer in the TMN model. Figure 7.3 depicts this agreed breakdown of technologies.

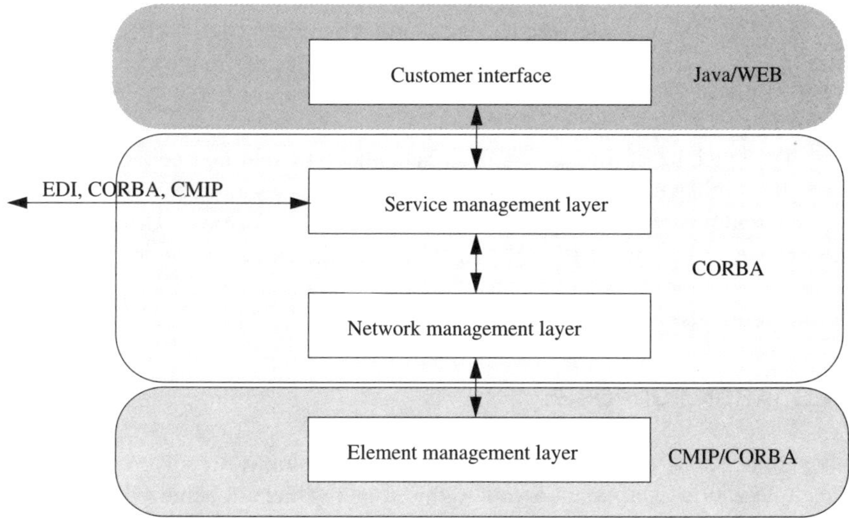

Figure 7.3 TMF strategic roundtable mixed technology architecture.

Java- and Web-based technologies were chosen for the client interface as this technology is perfectly suited to building lightweight GUI and easily integrates with distributed object-based technologies such as CORBA.

CORBA was viewed as the correct solution for the service management and network management layers of the TMN model. The CORBA services though not all implemented yet, provide the functionality that is required to build this type of software. The key features were improved scalability and location transparency which were seen as features lacking from the other alternatives. More importantly, CORBA is not seen as a replacement for other technologies such as OSI or SNMP. It is rather treated as an integration framework to provide great value-add to the existing technologies.

The element management layer was considered by most a problem that had been solved. While CMIP and GDMO were viewed as less than optimal solutions to the problem, the work that has been done by organizations such as ITU-T on MIBs was viewed as an adequate solution for this level. Some people expressed a desire to implement all new software solutions (i.e., mediation devices) using CORBA.

CMIP was viewed as the correct solution for the management of the network elements as most of the work had already been invested in this level of the TMN model. Effort defining the appropriate abstractions and models for the other layers (i.e., service management and network management layers) was viewed as a better investment of time and development effort.

The other technology options were chosen by equally small groups of about 10% each for both Java and CMIP.

The participants who chose CMIP justified its selection as the technology of choice based on its availability and its stability. While admitting that CMIP had known limitations and problems, they believed there were sufficient knowledge to enable CMIP to be a viable solution at all levels of the TMN model.

The people that supported the Java-based solution recognized the fact that they would require the services of an object request broker type architecture and thus they would require CORBA in order to provide a solution. They however chose Java on a development productivity-based argument which relies on the production of Java chips and other advancements in technology.

The interface used between pairwise service management domains was a cause of some debate as the common interface used in North America is EDI, while in Europe some implemented trouble-ticketing systems use CMIP. A number of people expressed a desired to use CORBA if the service management software within their own domain was implemented using CORBA.

The outcome of this meeting validates the view that no one technology will dominate the TMN architecture and that appropriate technologies should be chosen for each level. It is not important which technologies are chosen during this exercise, the candidate technologies will change over time. It is important however, that industry agrees with the notion of an integrated solution from multiple technologies and a commercially focused process to evaluate and define the integrated solutions.

7.7 INTEGRATION PURPOSES

Integration is one way of deploying multiple technologies to solve complex business problems. Integration itself is not a goal, rather it is a vehicle to achieve business goals. For two given technologies, there may exist several different ways to integrate them. It is often hard to reach a simple conclusion that one approach is the best before we know why we are trying to integrate them.

In the area of telecommunication network and service management, technology integration is required to achieve interoperability between different management domains. A large network management environment may use different management technologies such as CMIP and SNMP in different subsystems. An integration is thus required in order to support the communication between these subsystems.

The integration's goal is also to strengthen the existing solutions by introducing new technologies which offer better features in certain functional areas. For example, we introduce distributed technology to provide the management system with a distributed environment. This introduction requires the integration of the distributed technology with the existing management solutions.

It is also useful to provide a domain-independent (or protocol-independent) application development and deployment environment. In a large telecommunication network, it is common that different network equipment and network technologies (ATM, SDH/SONET, WDM, FrameReply, etc.) are used. They are normally managed by different systems using different management technologies (CMIP, TL1, SNMP, CORBA, Java). It is always desirable but hard to develop network management systems supporting transparent end-to-end applications from the individual underlying management technologies. The problem becomes more acute when the migration of these technologies takes place. In many cases, the migration forces the change in the management applications—a painful process that should be avoided at any cost. To make the management applications independent from the underlying technologies, an integration layer is required which encapsulates all the underlying technologies and provides a domain-independent environment for the applications.

The service management and network management have different but closely related business focuses. Telecommunication services are supported by networks and elements. The service management layer needs to communicate with the network management layer in order to perform its management activities. The management functions supported by these layers and the technologies used in managing these layers need to be integrated to provide a better service and network management environment.

Finally, technology integration also helps migrating the legacy systems to more advanced technology frameworks for application development and deployment. Integration is also a common mechanism to allow legacy systems to be used in a newly developed system to reduce the development cost.

In TMN network and service management, perhaps the most important purpose of the integration is to achieve the following two management scenarios:

- Domain (or protocol) independence managers and applications
- Service and network management integration

Figure 7.4 depicts a scenario where a network technology independent component is required to support more generic network and service management functions. This manager acts as a cross-domain manager which insulates the applications from the underlying network and element management systems. It also provides, through adapters as integration points, interworking with different legacy systems.

This scenario provides an integrated network management system supporting a truly end-to-end network view of TMN functions. An example is that the cross-domain manager is a network connection manager which provides end-to-end connections across all different subnetworks with different network technologies and management systems.

Another integration scenario is the integration between the service management and the network management. Lack of this integration is well recognized as the key problem inhibiting many service providers to operate smoothly.

Figure 7.5 illustrates three types of integrations relevant to TMN management:

- *The network interoperability*—This integration deals with the business requirements to integrate different network equipment and subnetworks of different technology. OSI initiative provided a framework to support this integration in one technology domain.
- *The network and service management integration*—This integration deals with the business requirements to support smooth information and process flow from services to network. Some work in the TINA-C service architecture and TMF SmartTMN initiatives provided technology solutions to support this integration. An explicit domain

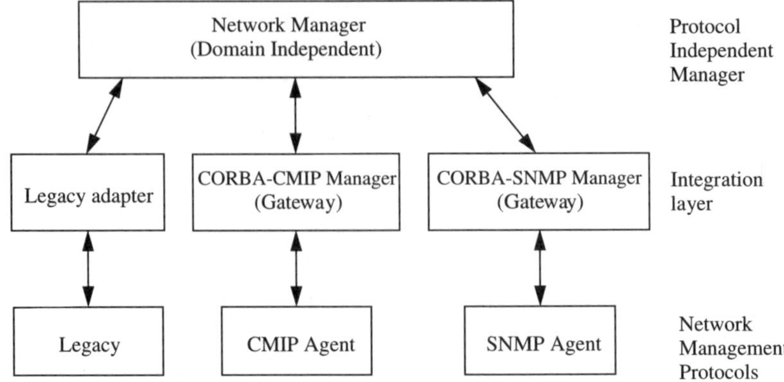

Figure 7.4 Cross domain manager.

independent layer, as depicted in Figure 7.4, provides a much better support for this integration.

- *The interworking between service management processes and domains*—This integration deals with the business requirements to support integrated services and management application development within a service provider or between service providers. The recent technology direction set by TMF (Telecom Operation Map and Technology Integration Map) provided technology solutions to support this integration.

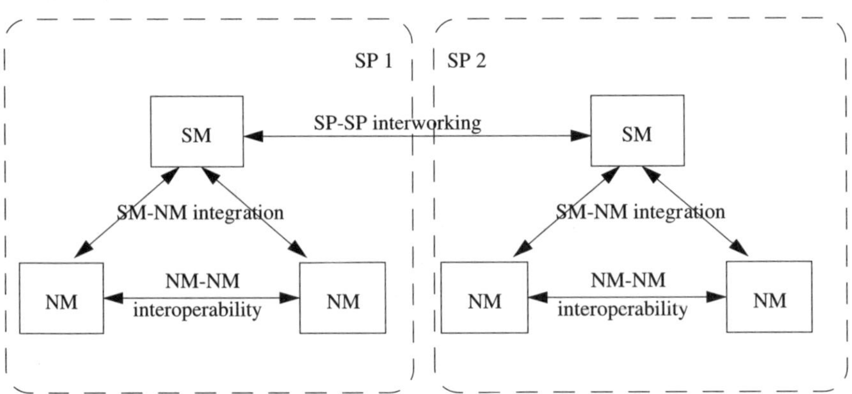

Figure 7.5 Intra-SP and inter-SP integrations.

In the following chapters, we focus our effort on these types of integrations. We discuss the different integration approaches, the process one should use to define generic information models and integration architectures.

7.8 SUMMARY

Whereas it is important and a sign of maturity of the industry to recognize and accept the notion of right technology for the right domain, it requires more substance to instantiate the notion: that is, what are the integration approaches and how to achieve an integrated information model and business process model? These issues remain to be major challenges to the telco industry.

8

Integration Approaches

It is quite observable that the religious wars on which technology is best in the network management area is becoming less of a focal point. The current industry trend is to recognize the multiple technology coexistence; to assess their values to the management solution; and to create better integrations between them.

The integration of OSI, CORBA, and Java/Internet to offer integrated TMN solutions is one of the most urgent issues in the industry. As discussed before, there are different approaches to integration, and each for a different purpose. In this chapter, we discuss issues related to technology integration by looking at some of these approaches in more detail. We do not intend to discuss all integration methods, rather we concentrate on the processes to achieve the integration. Given the prominence of the OSI technology in traditional TMN framework and the recent industry acceptance of CORBA, the integration between these two is used here as an example.

There are a number of approaches that can be used to support integrations between CORBA and OSI network management environments [43]. Three examples of such approaches are investigated here in more detail:

- Gateway approach
- Abstract object definition approach
- Application interworking approach

8.1 GATEWAY APPROACH

A *gateway* is a piece of software which sits in the intersection between two different environments with different object models. The gateway behaves like a mediation device to translate object definitions and functions from one environment to another. One way of interfacing CORBA- and OSI-based TMN systems is to build a gateway between the CORBA environment and the OSI network management environment. The main functionality of the gateway is to perform the protocol translation between systems developed in these two different technologies.

There are two typical scenarios of using such a gateway in CORBA/OSI network management systems:

- CORBA manager and CMIP agent
- CMIP manager and CORBA agent

If one wants to build a CORBA-based TMN management system which requires interaction with an existing OSI CMIP agent, a CORBA manager can be developed for this purpose. However, in order for this manager to communicate with the agent, an integration gateway is required. This gateway helps the manager communicate with the CMIP agent in order to carry out the management activities. This is the scenario of a CORBA manager interacting with a CMIP agent.

For such a system, a GDMO-based object model has already been defined on the agent side. For each network resource to be managed, a managed object is defined and instances of it are stored in the MIB. The CMIP protocol is used by the agent as the only communication protocol to receive management requests from a manager and to send event notifications to a manager.

On the manager side, CORBA IDL is used for object interface definitions and the communication is made through an ORB using IIOP.

A gateway is thus required between these two different systems. On the one side, it communicates with the CORBA manager using IIOP. On the other side, it communicates with the agent using CMIP protocol. A protocol translation is performed for the communication in both directions.

The system architecture of CORBA manager–CMIP agent gateway approach is depicted in the Figure 8.1.

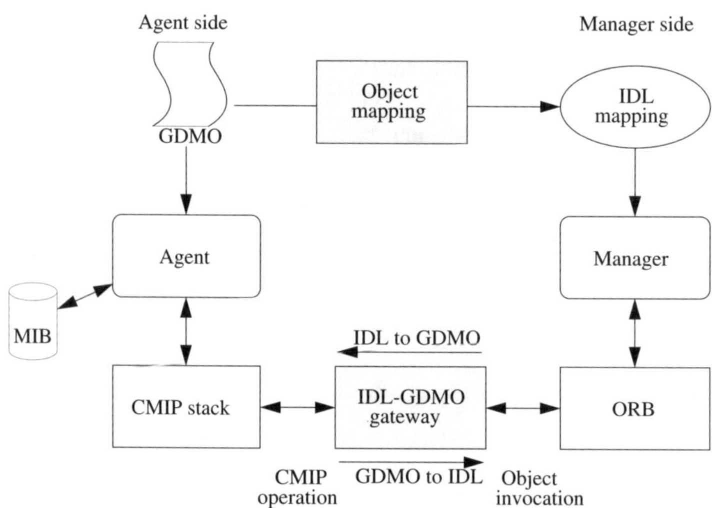

Figure 8.1 Gateway between a CORBA manager and a CMIP agent.

In some occasions, there are needs to develop new CORBA agents (or subnetwork management systems) which are required to interface upwards with the existing OSI systems. In this case, the agents in CORBA environment and managers in OSI environment should interwork. This scenario is not as common as the CORBA manager–CMIP agent scenario. This is because there are more requirements to build new CORBA managers to communicate with the existing CMIP agents.

In CORBA agent–CMIP manager scenario, the manager side uses a GDMO-based object model and the agent side uses IDL-based object interface definitions. The communication protocol used by the manager is CMIP and the communication used by the agent is a CORBA-based object invocation. Similar to the CORBA manager–CMIP agent scenario, a gateway is required to perform the protocol translation between the manager and the agent.

These two scenarios cover the most applications of the CORBA–CMIP gateway in an integrated system.

8.1.1 Integration Issues

In order to achieve communications between a manager and an agent for the purpose of exchange management information and functions, agreements must be made at two levels:

- *At the object definition level*—The development of the OSI-based TMN network management so far, in particular the development of TMN standards, has resulted in many GDMO object specifications. This has further led to the development and deployment of many management applications based on these object specifications. In the case of a CORBA–CMIP gateway, each of these GDMO objects needs to be mapped into an IDL interface object in the CORBA space. These mappings include the object definitions (such as attribute types and values), operations that can be performed on the objects (such as get, set, create, delete, action, cancel get), notifications emitted by the objects, and the behaviors of the object. The extent of these object mappings is illustrated in the Figure 8.2.

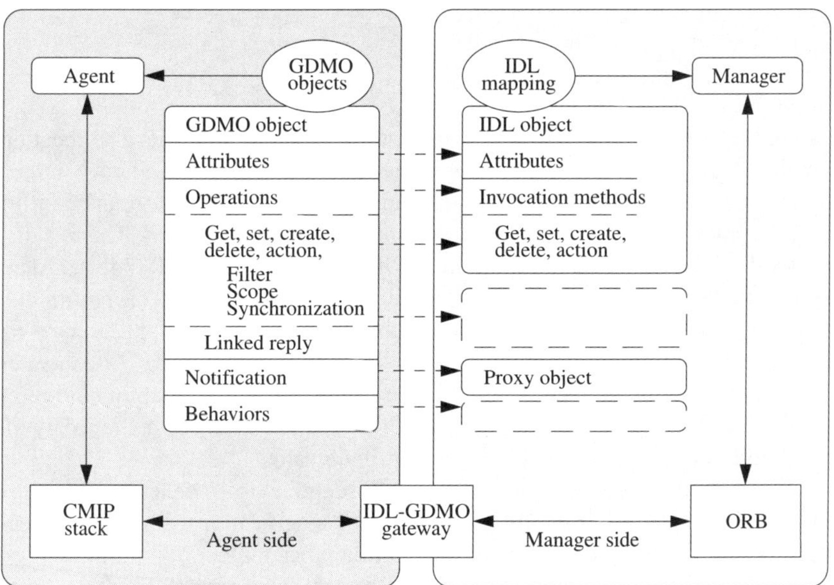

Figure 8.2 Object mappings in the gateway approach.

- *At the communication protocol level*—Operation requests sent by a manager using the *manager protocol* (i.e., CORBA manager using IIOP and OSI manager using CMIP) need to be translated at the gateway into the *agent protocol* before being passed to the agent. Operation responses sent by the agent using the agent protocol need to be translated back to the manager protocol at the gateway before being passed to the manager.

  Notifications issued by an agent using the agent protocol need to be translated at the gateway into the manager protocol before being passed to the manager. Confirmations of the notifications sent by the manager using the manager protocol need to be translated back to the agent protocol at the gateway before being passed to the agent.

There are also different ways to deploy the gateway. When the manager and the agent are distributed across different machines, there is an issue of where the gateway should be located. The location of the gateway affects the total system performance. Using CORBA manager and CMIP agent as an example, Figure 8.3 depicts three different possible locations.

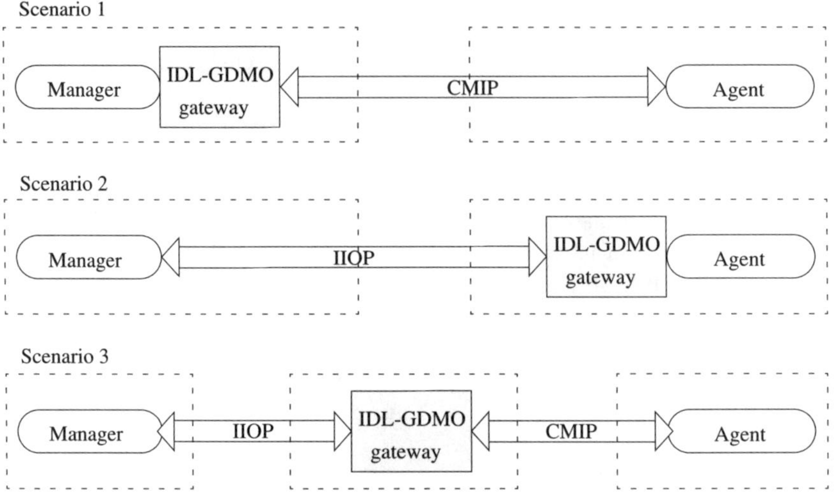

Figure 8.3 Location of integration gateway.

In Figure 8.3, each dotted box indicates a separate machine. The location of the gateway determines which protocol is used for remote communication between the manager and the agent. The different deployment strategies determine the performance influences.

For each GDMO object, an equivalent IDL object is defined. This results in a conceptual level of one-to-one mapping between GDMO objects and IDL objects. Also, for each operation in a GDMO object, an equivalent method in IDL definition is required. Statically, the mapping between GDMO operations and IDL object invocations appears one-to-one. For example, each `get` operation on an attribute defined in GDMO has an equivalent `get_attribute` operation defined in IDL. Dynamically, one CMIP `get` operation may fetch values of multiple attributes of a base object. In CORBA, this can be achieved by multiple object invocations; each invocation returns one attribute value.

Commercial tools of various degree of strength exist to generate IDL object definitions automatically from original OSI GDMO object specifications. These will help develop and deploy ORB–CMIP gateways in management applications.

In addition to the object mapping and protocol translation, there are many other complex issues which have to be addressed when a gateway is used to integrate these two technologies. The complexity is caused by the fact that CORBA and CMIP are representatives of two different management paradigms. Although both are object oriented, there are still significant differences between them as discussed in Chapter 6. Table 8.1 summarizes the major differences.

In essence, a gateway approach is an approach to mediate two object models to reduce the differences listed in the table below. In the gateway approach discussed in the next section, we show how this approach addresses some of these issues.

8.1.2 Example of the Gateway Approach

Efforts have been made by a number of organizations on the gateway-based integration approach. Probably the most noticeable is the work of the *Joint Inter-Domain Management (JIDM)* working group, which is supported by The Open Group and the TeleManagement

TABLE 8.1 Differences between CORBA and CMIP Objects

Features	OSI	CORBA
Communication	Between management entities	Directly between objects
Operation mode	Asynchronous operation requests and responses	Synchronized RPC-based object invocation
Object instances	Stored in a tree structured MIB managed by an agent	Object implementations
Object encapsulation	All attributes are visible, both object level operations and attribute level operations are defined	Attributes are encapsulated, object invocation is at object level
Operations on multiple objects (Bulk operation)	Supported by scope and filtering as well as linked replies	Not directly supported
Object identifier	Distinguished name (DN) includes navigational information, location is not transparent	Object reference which is location transparent
Events	Notifications supported by asynchronous communication	Using pull or push event channel

Forum. The JIDM specifications include three types of gateway definitions: GDMO to IDL, IDL to GDMO, and SNMP MIB to IDL. Each type deals with its own special technical challenges. However, the basic principles behind them are the same. In the remaining text of this section, we use the GDMO to IDL gateway as the example to demonstrate the basic functionality and characteristics of the JIDM gateway approach.

The JIDM's GDMO to IDL gateway specification contains two major parts. The first part is the *Specification Translation* which focuses on the object conversion and protocol translation between GDMO object model and IDL interface objects. The second part is the *Interaction Translation* which focuses on the interaction process between two different domains. It addresses most of the issues summarized in Table 8.1, such as, the mapping of object identifiers between GDMO DN and CORBA object reference and the way of handling CMIP scoping and filtering.

8.1.2.1 Specification Translation. The Specification Translation [103] stipulates a number of translation rules between two object models. These rules are concerned with two issues. The first issue is how to translate a GDMO object class template into a CORBA IDL object interface; and the second issue is how to translate the ASN.1 definition of the CMIP protocol into a set of CORBA IDL definitions.

The GDMO object class template contains the following components:

- Attribute package template is a group of attributes used to address one functional area. Attribute level operations (standard CMIS `get` and `set`) are associated with each attribute in the package.
- Notifications template is used to define types of notifications which can be generated by an object.
- Action template is used to specify any action that is not supported by standard CMIS operations.

For GDMO object templates to CORBA IDL translation, the translation algorithm is applied to each template recursively, starting from the `MANAGED OBJECT CLASS` template. The GDMO to IDL translation has the following features:

- In CORBA, an IDL interface `ManagedObject` is defined, which inherits all properties from an IDL interface, named `Object`. All interfaces generated from the translation of GDMO objects inherit from this `ManagedObject` interface. All the common functionality to support CMIS requests and responses is provided through the implementation of the `ManagedObject` interface.

- A managed object class defined in GDMO is mapped to an IDL interface with the same name. Some additional interfaces to support notifications are required.
- Attributes in each package are translated as a set of IDL operations based on the attribute property as indicated in the Table 8.2.
- Each parameter is mapped to an IDL type.
- Actions are mapped to IDL operations.
- Notifications are mapped to IDL operations. There are two types of notification interfaces, `Notification` and `PullNotification`. These two types of operations are associated with the two event service modes supported by CORBA event service (*push* mode and *pull* mode).

Table 8.2 lists the translation of CMIS operations applied to each attribute according to the GDMO definition.

TABLE 8.2 Mapping Attribute Properties to IDL Operations

GDMO Attribute Property	IDL Operations Required
GET	<attribute-label>Get
REPLACE	<attribute-label>Set
GET-REPLACE	<attribute-label>Get and <attribute-label>Set
ADD	<attribute-label>Add
REMOVE	<attribute-label>Remove
ADD-REMOVE	<attribute-label>Add and <attribute-label>Remove
REPLACE-WITH-DEFAULT	<attribute-label>SetDefault

The second issue the specification translation is concerned with is how to translate ASN.1 definitions of the CMIP protocol into a set of CORBA IDL definitions. This translation focuses on the data type translations which cover both base ASN.1 data types, such as `INTEGER`, `OCTEC STRING`, `ANY`, and constructed types using these based types.

For ASN.1 to IDL translation, rules are specified to translate each ASN.1 data type to a corresponding IDL type. Using CMIP `get` operation as an example. The main parameter for a CMIP `get` operation is `GetArgument`. Figure 8.4 shows the ASN.1 definition for the

```
GetArgument ::= SEQUENCE {
   COMPONENTS OF BaseManagedObjectId,
   accessControl    [5] AccessControl OPTIONAL,
   synchronization  [6] IMPLICIT CMISSync DEFAULT
                         bestEffort,
   scope            [7] Scope DEFAULT baseObject,
   filter           CMISFilter DEFAULT and {},
   attributeIdList  [12] IMPLICIT SET OF AttributeId
                         OPTIONAL
   }
```

Figure 8.4 ASN.1 definition of `GetArgument`.

GetArgument in GDMO. This data type is defined by a constructed type SEQUENCE, each field in the SEQUENCE has a name of the field and its data type. Figure 8.5 shows the resulting IDL definition of GetArgument after applying the translation algorithm.

```
struct GetArgumentType {
    ObjectClassType baseManagedObjectClass;
    ObjectInstanceType baseManagedObjectInstance;
    AccessControlTypeOpt accessControl;
    CMISSyncTypeDef synchronization;
    ScopeTypeDef scope;
    CMISFilterTypeDef filter;
    GetArgumentAttributeIdListTypeOpt attributeIdList;
};
```

Figure 8.5 IDL translation of GetArgument type.

For each field in the ASN.1 definition of GetArgument, an equivalent IDL data type is defined. For each constructed data type used in the GetArgument definition, a translation is required to yield its equivalent IDL definition. For example, the first field in the ASN.1 GetArgument is baseManagedObjectId, Figure 8.6 shows its original ASN.1 definition, and Figure 8.7 shows the resulting IDL definition after the translation.

```
BaseManagedObjectId ::= SEQUENCE {
    baseManagedObjectClass ObjectClass,
    baseManagedObjectInstance ObjectInstance
}
ObjectClass ::= CHOICE {
    globalForm [0] IMPLICIT OBJECT IDENTIFIER,
    localForm  [1] IMPLICIT INTEGER
}
ObjectInstance ::= CHOICE {
    distingusiedName          [2] IMPLICIT DistinguishedName,
    nonSpecificForm           [3] IMPLICIT OCTEC STRING,
    localDistinguishedName [4] IMPLICIT RDNSequence
}
```

Figure 8.6 GDMO definition for BaseManagedObjectID.

This translation algorithm needs to be applied recursively to all constructed types until the base data types, such as INTEGER and OCTEC STRING, are encountered. Each ASN.1 base type is then translated into its equivalent one defined in CORBA IDL.

8.1.2.2 Interaction Translation. The Interaction Translation [91] contains a set of CORBA facilities required to support the interworking between CORBA and CMIP management environments. The reason for having this part of the translation is to solve the problems caused by the technology differences between CORBA and CMIP listed in Table 8.1. Let's see how some of these problems are solved in JIDM interaction translation specification.

In OSI, communication is carried out between a pair of management entities, i.e., between a management entity playing a manager role and another entity playing an agent role. An agent is in control of all the object instances stored in its MIB. An operation to a managed object controlled by the agent is invoked on the agent. In CORBA, object invocation is

```
enum ObjectClassTypeChoice {
   globalFormChoice_3,
   localFormChoice_3
};
union ObjectClassType
   switch(ObjectClassTypeChoice) {
         case globalFormChoice_3:
               ASN1_ObjectIdentifier globalForm;
         case localFormChoice_3:
               ASN1_Integer localForm;
};

enum ObjectInstanceTypeChoice {
   distinguishedNameChoice,
   nonSpecificFormChoice,
   localDistinguishedNameChoice
};
union ObjectInstanceType
   switch(ObjectInstanceTypeChoice) {
         case distinguishedNameChoice:
               DistinguishedNameType distinguishedName;
         case nonSpecificFormChoice:
               ASN1_OctetString nonSpecificForm;
         case localDistinguishedNameChoice:
               RDNSequenceType localDistinguishedName;
};
```

Figure 8.7 IDL translation of `BaseManagedObjectID`.

made to an object implementation. The management entity concept is not directly supported by CORBA. JIDM interaction translation specification is an attempt to support this facility. It defines the following concepts:

A **manager domain** is a group of entities playing manager roles in management activities between open systems. A manager domain is sometimes also referred to as a manager, or a manager application.

A **managed domain** is a group of entities playing agent roles in management activities between open systems. A managed domain is also referred to as an agent or a managed object domain.

With the support of the manager and managed domains, the management functions are performed by exchanging management requests through these domains, rather than directly invoke methods on managed objects. That is, these domains correspond to the concept of management entity in OSI.

In the interaction translation, JIMD uses the CORBA facilities and CORBA interfaces to achieve interworking with other non-CORBA management environment. These facilities and interfaces are defined at three different levels:

- *JIDM facilities*—this is a logical group of all the facilities which are required to support access to a manager domain and managed domain. These facilities cover the common management functions defined in the OSI or SNMP management reference model. Examples of such facilities and services include object creation, operation invocation, event reporting, and distribution.

- *OSI management facilities*[1]—This is a logical group of all the facilities and interfaces required to support interworking with an OSI management environment. This group of management facilities provide a CORBA view of the OSI management reference model. It extends the generic JIDM facilities to support CMIS interactions in CORBA, such as `cmis_get`, `cmis_set`, `cmis_create`, `cmis_delete`, etc. It also supports OSI-specific concepts, such as scoping, filtering, and linked replies.
- *Specific interfaces*—This is a group of interfaces with specific functionality required to interwork with OSI information model. The OSI management facilities are reused and extended by these interfaces. For each object model defined in GDMO/ASN.1, an equivalent CORBA IDL object model can be generated by following the translation rules specified in the specification translation.

To understand how the JIDM interaction translation works (in conjunction with specification translation), let us look closely at two technical aspects of the JIDM approach:

- Communication between manager and agent
- Scope and filtered operations

First let's look at the communications between management entities. In OSI, management activities include:

- Managers send operation requests to agents for object creation, deletion, get attribute value, or set attribute value.
- Agents send events to managers to inform the change of object states.

The management environments targeted by JIDM (both OSI and SNMP) all use manager–agent communication paradigm. Therefore, the support for manager–agent communication forms a core part of JIDM interaction translation.

In the scenario to integrate a CORBA manager with an OSI agent, in order for a CORBA manager to create a managed object through the OSI agent or to perform operations on the OSI agent, a `ProxyAgent` and a number of supporting interfaces are defined. An instance of the `ProxyAgent` object is created when a manager gains access to an agent. This object contains operations which can be used to obtain references to a suitable factory through which the new object is to be created.

In order to support a CORBA manager to receive events from an OSI agent, two interfaces: `EventPort` and `EventPortFactory` are defined on the manager side. Access to a manager is gained through an `EventPort` object.

In the scenario for an OSI manager to communicate with a CORBA agent, a `DomainPort` (and `DomainPortFactory`) interface is defined in CORBA agent. This interface controls all access made by the ISO manager.

To enable a CORBA agent to forward events to OSI managers, an `EventPortFinder` interface is defined. This interface is used to determine the destination of event forwarding operation based on the specification of a set of `Keys` and `Criteria`. The functionality of this interface is very similar to that of `EventForwarding Descriminator` defined in OSI (ITU-T X.721).

The following example shows how a CORBA manager creates a new managed object in an agent. The same example also shows how the CORBA agent responds in the scenario where CORBA is the agent. Figure 8.8 depicts this object creation example.

[1]An equivalent set of SNMP Management Facilities is also defined for CORBA/SNMP integration.

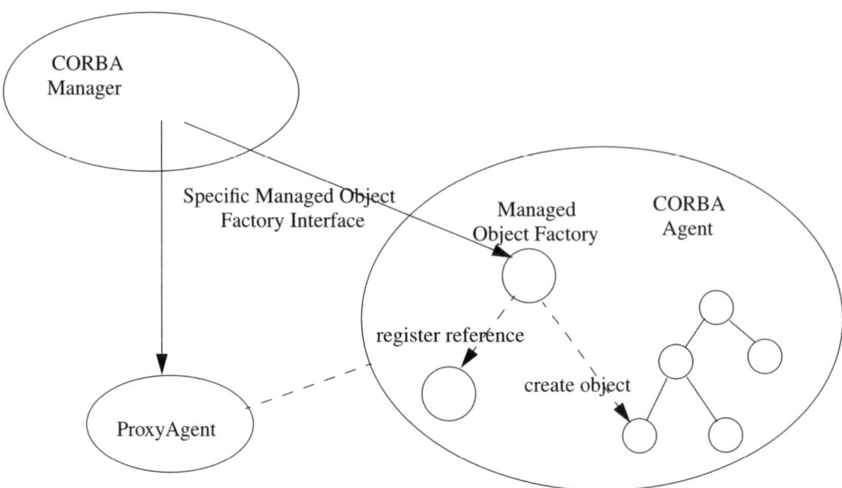

Figure 8.8 Creating a new managed object through CORBA.

To create an object, the CORBA manager needs to perform the following activities:

1. First, it communicates to the `ProxyAgent` acting as a representative of the management agent in the CORBA environment.
2. The `ProxyAgent` provides the required information to the CORBA manager. This information enables the manager to find a suitable managed object factory for object creation.
3. The CORBA manager then invokes a suitable operation on the managed object factory to create the new managed object.

In the scenario where CORBA is playing the agent role, when receiving the object creation request from a manager, the CORBA agent needs to perform the following actions:

1. The CORBA agent, through managed object factory, creates the CORBA managed object and performs the following actions:
   –register a reference to the managed object with a name in the local naming context,
   –notify registered managers that a new managed object has been created.
2. If the creation operation is successful, the managed object factory returns a reference to the manager who invoked the operation. Otherwise, it raises an exception.

The other issue the JIDM interaction translation addresses is OSI scope and filtered operations as well as the agent's ability to send linked replies.

CORBA cannot bind a tree structure to its objects. So to perform an operation on a tree structured object hierarchy becomes an issue when CORBA is used to closely map OSI functions.

The JIDM interaction translation addresses this issue in the OSI management facilities specification. To support scoped operations, the following capabilities must be supported:

- ability to send scoped operations—a requirement to a CORBA manager
- ability to performed scoped operations—a requirement to a CORBA agent
- ability to send bulk results—a requirement to a CORBA agent
- ability to receive bulk results—a requirement to a CORBA manager

In JIDM interaction translation, four scoped operations, `scoped_get`, `scoped_set`, `scoped_action`, and `scoped_delete`, are defined as part of the `ManagedObject` interface. This interface allows a manager to send a scoped operation explicitly. As part of each scoped operation, two parameters are provided: a reference to a `LinkedReplyHandler` object and a reference to an `EndofRepliesHandler` object. The first reference allows the manager to receive linked replies for the scoped operation whenever the `send_reply` method (of the `LinkedReplyHandler` class) is invoked. The second reference allows the method `end_of_reply` (of `EndofRepliesHandler` class) to be invoked to signify the end of linked replies.

The `LinkedReplyHandler` object, in conjunction with the `ProxyAgent` object, can be used to support asynchronous mode of operation requests.

On the CORBA agent side, the ability to perform scoped operation relies on the base managed object to propagate the requests to its descendants. This implies each managed object in CORBA must keep the information of all its descendants. In other words, the virtual MIB tree structure must exist.

In addition, when a CORBA managed object is involved in a scoped operation, it must propagate management requests to each of its descendants. Each descendant will send the operation reply to the manager directly, but will notify its superior the end of its operation in a recursive manner. When the base object receives all such end of operation notifications, it will invoke the `end_of_reply` method to notify the manager.

The following example in Figure 8.9 describe how a CORBA manager invokes a scoped and filtered operation. The same example also shows how CORBA agent responds when it is used as an agent.

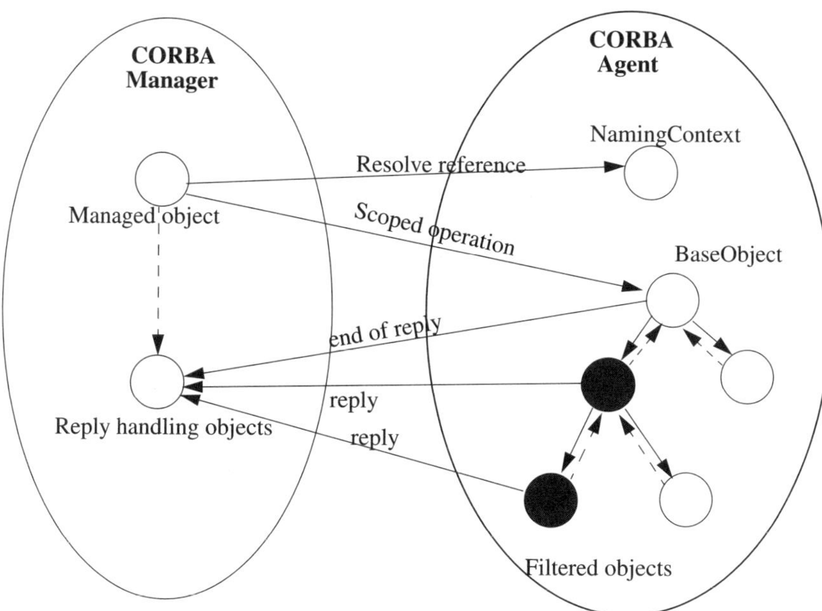

Figure 8.9 Invoking scoped and filtered operation.

In order to perform a scoped and filtered operation, the CORBA manager needs to perform the following actions:

1. The CORBA manager must resolve the object reference of the base managed object in the scoped operation.

2. The CORBA manager selects a predefined object that has the capability to handle multiple replies. This object is used as the reply handling object. The reference of this object is passed to the agent with a scoped operation.
3. The CORBA manager invokes the appropriate scoped operation (`scoped_get`, `scoped_set`, `scoped_action` or `scoped_delete`) with the reference to the reply handling object.

In the scenario where CORBA is playing the agent role, when receiving the scoped operation request from a manager, the CORBA agent needs to perform the following actions:

1. The base managed object propagates the request to its descendants with the reference of the reply handling object supplied by the manager. This requires each managed object to have knowledge of its descendants.
2. Each managed object within the defined scope and satisfying the filter sends the reply to the requested operation to the reply handling object by invoking its `send_reply` operation.
3. The base managed object is notified when all descendants contained in the scope complete their operation.
4. If the whole operation is successful, the base managed object invokes the `end_of_replies` operation of the reply handling object. Otherwise, it raises an exception.

By introducing these new facilities to closely map the OSI concepts and features, the JIDM approach makes it easier to support CORBA/OSI gateway.

8.2 ABSTRACT OBJECT DEFINITION

Different from the one-to-one object mapping based gateway approach, the Abstract Object Definition approach (or Abstract Mapping approach) addresses the integration problem from a different viewpoint [43].

In many real-world object modeling exercises, different object granularity is used for different modeling concepts at different levels of abstraction. For example, when modeling a network element, such as a switch, all the details of the switch need to be modeled which may include cards, ports, and other internal details. These details are useful for operations related to the management of that switch. When modeling a subnetwork, the internal details of each individual switch are no longer important. Instead, only the access points of switches and the connections between switches are important (these form the so-called the network view of network elements). When modeling a global network containing many subnetworks, the features of the subnetworks, such as the types of subnetwork (in terms of the transport protocol supported by the subnetworks, e.g., ATM and SDH), the termination points of each subnetwork, the connections between subnetworks are important. The internal connections within a subnetwork become less important for a global network manager.

These observations state an obvious concept that in a layered network structure it is important to allow different levels of object abstraction to be defined for different layers. This is particularly important to support the management of a large-scale network. The object specification at a higher level should be viewed as a simplification and generalization (with more general concepts) of the object specifications at the lower level. The object specifications at the lower level should be viewed as the specialization and the instantiation (with more details and specifics) of the objects at the higher level.

This is also true for the differences between the TMN service management layer and network management layer. The service management layer is more interested in abstract services supported by underlying networks while the network management layer is more interested in network details which form the basis for the services. Thus, each higher TMN management layer is seen as a further object abstraction and simplification of the previous layer.

One of the common criticisms of the TMN model is its complexity, in particular with its different layers. However, the TMN model should not be viewed as a complex architecture if the layers of TMN are regarded as the different viewpoints of the same object hierarchy (i.e., the business view, service view, and network view). Each different viewpoint defines a simplification of the previous viewpoint and supports a different set of operations and services. These different viewpoints (or layers) should not necessarily be regarded as different layers of software stacked together. Instead, they reflect the different access points in the whole object hierarchy (which may not exist explicitly in its entirety), as depicted in the Figure 8.10.

The abstract object definition approach is based on this principle. It treats each level of object abstraction as an integration point between different management technologies.

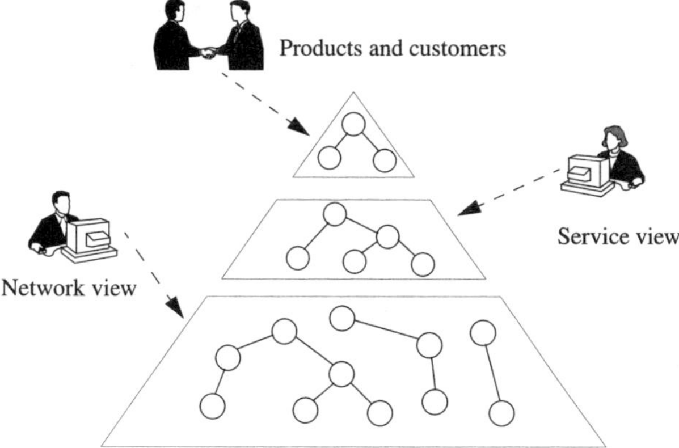

Figure 8.10 Object abstraction of TMN.

8.2.1 General Approach

The abstract object definition approach has two specific design goals:

- The ability to manage a large network at a higher abstraction level with more abstract concepts
- The ability to retrieve details when required

This approach introduces a set of CORBA objects at an appropriate level of the management hierarchy. These CORBA objects represent a view more abstract than the underlying OSI system required for integration. The underlying OSI system, represented by a set of GDMO objects, is viewed as the specialization of the CORBA objects. The GDMO objects, are therefore mapped into the CORBA objects, but not necessarily in a one-to-one manner.

For example, a set of TMN service objects can be defined using CORBA IDL. These services are implemented on top of a set of GDMO objects representing network elements. The relationships between the IDL service objects and the GDMO network element objects can be specified as part of the object modeling for the TMN management hierarchy. These relationships reflect the rules of the object specialization and generalization. The implementation of these rules forms the integration gateway.

Each object invocation on a CORBA object is mapped into a set of CMIP operations performed on a set of GDMO objects according to the relationships specified between them.

For example, a service activation request can be mapped onto a set of configuration or creation operations on a set of network element objects.

Events occurring on a GDMO object can be reflected at the CORBA object level. For example, when an alarm is generated by a network element, the relationship rules can be used to determine which services are affected and to what level. This allows service level management tasks to be performed, such as informing the service customers about the degradation in the quality of a service.

CORBA and CMIP can be used together to manage telecommunication networks and services. For example, OSI CMIP-based communication protocol can be used for network management related communications between a manager and an agent. CORBA communication protocol can be used for service management related communications with service objects defined in CORBA IDL.

Figure 8.11 depicts the integrated TMN management environment where the abstract object definition approach is used. In this scenario, a set of CORBA objects, abstract network objects, are defined to offer higher level network management functions to directly support service objects. The modeling of these objects is an integral part of the service object modeling. These objects are implemented by using OSI communication to interact with OSI GDMO objects.

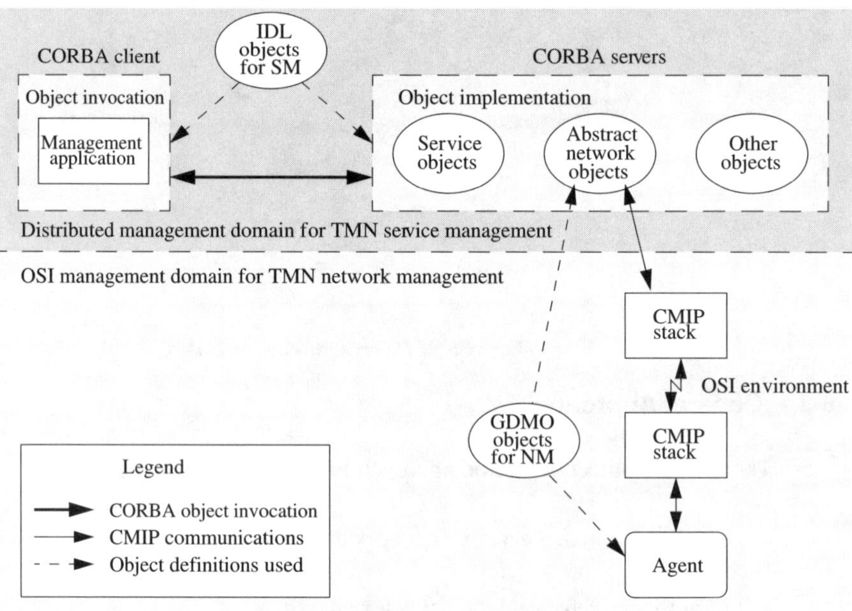

Figure 8.11 Integrated management environment.

The abstract object definition approach does not exclude one-to-one object mapping between certain objects. For example, at the service management layer, a service access point may well be one-to-one mapped onto a network termination point at the network management layer. However, the services supported by the service access points and the network termination points can still be quite different. The former is part of the service object modeling concept, while the latter is a network technology specific concept.

8.2.2 Characteristics

This approach is not based on one-to-one mappings between GDMO objects and CORBA objects. GDMO operations, such as get, set with filter and scope conditions, are not mapped in the CORBA object environment.

The main characteristic of this approach is that the CORBA object modeling will be achieved strictly from the business requirements for which CORBA is selected to use. This object modeling exercise is independent from the underlying GDMO object model used to support the CORBA layer and is independent from the underlying management technology support. For instance, a broadband connection can be sufficiently modeled as a path between two end points as far as the end-to-end service provisioning function is concerned. Further down in the provisioning chain, however, there is a need to map this connection into, say, an SDH trial in the GDMO world, which further defines multiple types of termination points, link connections, subnetwork connections and cross connections, etc. A set connection operation performed on the CORBA object may result in a sequence of `set` operations in the GDMO world, and some of the `set` operations may require scope and filter conditions. However, these conditions are not visible to the CORBA object. As far as the CORBA object modeling is concerned, the *semantics* of the create connection interface are mapped into a collection of operations performed on a set of GDMO objects. There is clearly not any specific one-to-one mapping between them.

This is the principle of using GDMO objects and CMIP operations, using to support a higher level and more abstract object model expressed in CORBA. The (specific) relationships between these two object models are defined as part of the CORBA object modeling (indeed, this generalization/specialization relationship must be defined in any object modeling exercise). The implementation of these relationships forms a CORBA/CMIP integration gateway, which provides the glue between two operation systems.

This approach enables us to best use both CORBA and CMIP technology by using CORBA for service management and CMIP for network management. All service concepts (such as video on demand service, customer, account, etc.) and high-level network management concepts (such as video connection) can be represented as CORBA objects. The operations on these objects (such as creating a video connection, generating a bill) can be defined as CORBA methods. The semantics of the CORBA object model and operations are defined in terms of the underlying network management object model and the support system. These include the network configuration, CMIP managers and agents responsible for managing broadband networks, and VOD servers. The implementation of this semantics is indeed a gateway integrating a piece of service management functions (VOD management, cross-domain connection management) and the network management functions (domain specific connection management, VOD server management, etc.).

As far as the integration between TMN service management layer and network management layer is concerned, the key to the success of using this approach is the *properly* defined high-level TMN service objects. These object models have not yet been defined and agreed by industry and standards bodies. These object models play a vitally important role in achieving service level interoperability. Various standard bodies and technology forums have the programs to achieve these. One example is the TeleManagement Forum's SmartTMN program which aims to define a common object model and business process flow-through for an integrated TMN service and business management system. In the next chapter, we particularly review some of the work in the object modeling area.

8.2.3 Example

As stated before, the main characteristic of this approach is to properly model the CORBA environment, independent from the underlying GDMO environment. After that, all relationships between the CORBA objects and the GDMO objects need to be modeled. These relationships form the integration gateway. Therefore, the integration approach is largely a set of processes to be applied to individual integration requirements, rather than a generic approach which could result in an algorithm for automatic translation. To demonstrate the process, an example of CORBA/CMIP integration using the abstract object definition approach is given here. The example addresses the service provisioning of Video-On-Demand (VOD) services. Given the theme of this chapter, the example is considerably simplified to focus on the integration processes of this approach.

8.2.3.1 Scenarios. The VOD service provisioning scenario is illustrated in Figure 8.12. A number of customers and a number of VOD servers are connected through a network. Customers access VOD services by using the *video selectors* which send the commands, such as select, start, stop, pause, change channel, etc., to the VOD servers through the network. VOD servers perform actions according to the commands.

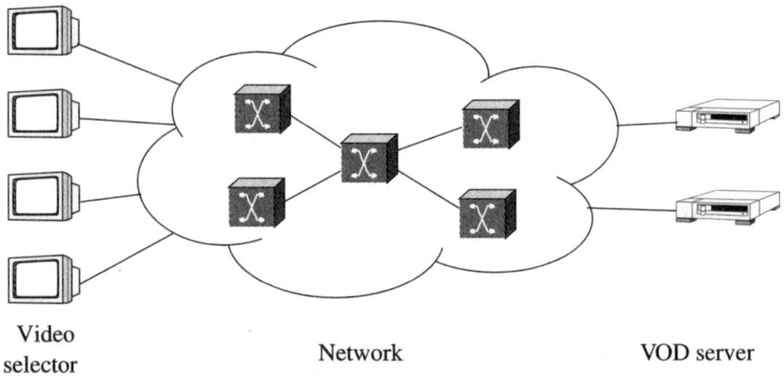

Figure 8.12 VOD service provisioning scenario.

The integration of CORBA/CMIP is required by the VOD service provisioning system for the implementation and management of the network-based services. The VOD service provisioning system is modeled and implemented as a CORBA subsystem, and the VOD servers and network managers are implemented in CMIP.

Different aspects of the CORBA/CMIP integration can be demonstrated using this example. For instance, both the abstract object definition approach and the IDL-GDMO gateway (one-to-one mapping) approach can be adopted at different integration points between CORBA and CMIP in the system.

8.2.3.2 Modeling. First, let's consider the modeling of this provisioning system. The following list of entities is modeled in the system:

- *VOD service session*—Representing VOD service offering. This object represents an instance of the interactions between a customer and service provider.
- *VOD connection*—Representing the end-to-end connection on which the video is transmitted.
- *VOD manager*—Representing the video transmission service.
- *Video selector*—Modeling the physical box-top selectors customers use to order video and control the operations.
- *ATM network*—Representing the physical network connections in, say, ATM network. Further detailed representation of the components of the ATM network (such as the ATM Forum M4 concepts) are omitted from the modeling.
- *VOD server*—Modeling the physical video transmission equipment.

Figure 8.13 presents an object instance hierarchy of the system. This object hierarchy supports the functional abstraction of VOD provisioning system from the network equipment to the customers access.

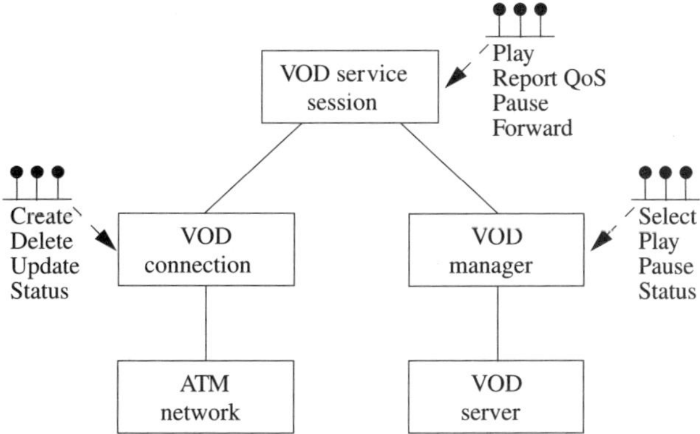

Figure 8.13 Object modeling of VOD service provisioning.

8.2.3.3 Architecture. The VOD service provisioning system contains three sections: a CORBA section, an OSI section, and a CORBA/OSI integration section.

The CORBA section consists of a number of CORBA objects. These objects act as service managers and are responsible for providing the VOD services and the management of this service.

The OSI section consists of a number of CMIP management entities. These entities manage network resources involved in providing a VOD service, such as the video selectors, the ATM network, and the VOD servers.

The CORBA/CMIP section consists of a number of integration objects which understand both CORBA service definitions and OSI network definitions. More importantly, these integration objects have the ability to perform the mappings between a CORBA service request and a set of OSI network management requests.

The architecture of the VOD service provisioning system is illustrated in Figure 8.14. It has following CMIP agents:

- ***Video selector***—The video selector is a CMIP agent which simulates the behavior of video selectors at the VOD customer premises.
- ***Network manager***—The network manager is a CMIP management entity. It plays both manager role and agent role. As a manager, it manages the ATM network supporting the network connections. As an agent, it receives connection requests from network users.
- ***VOD server***—The VOD server is a CMIP agent that simulates the behavior of a VOD server equipment.

The VOD service provisioning system has following CORBA objects:

- *Session manager factory*—The session manager factory is used to create session managers. It is a standard CORBA factory providing life-cycle management for CORBA objects. There is one session manager factory for each installation (host).
- *Session manager*—A session manager is responsible for managing a VOD service session. It receives events from the video selector (via the session managers selector) and coordinates the control of the VOD server and the connection manager.

The VOD service provisioning system has following integration objects:

- *Session manager selector*—The session manager selector captures the events generated by the video selector. The session manager selector creates session managers

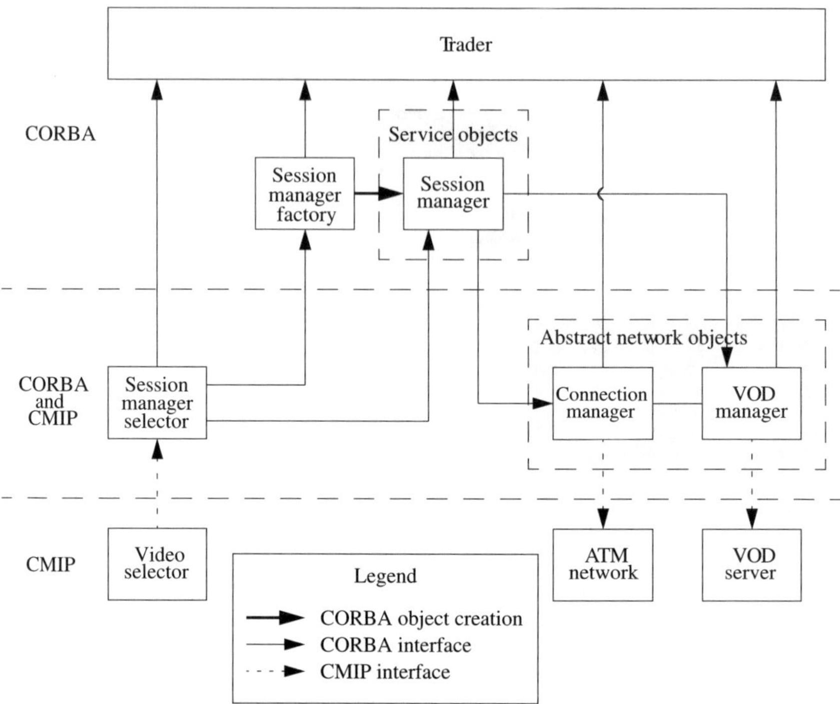

Figure 8.14 Architecture of the VOD demonstrator.

when required and forwards events from the video selector to the session manager managing that video selector.

- *VOD manager*—The VOD manager manages VOD servers. It answers the VOD requests from session manager by issuing CMIP requests to a VOD server.
- *Connection manager*—The connection manager is a CORBA object which manages the network connections. It provides an abstract, end-to-end view of the network and network services to the CORBA session manager.

The connection manager provides a set of CORBA interfaces for the session manager to set up, modify, and delete network connections. The implementation of these interfaces uses the CMIP interface to invoke ATM network functions to configure connections between the video selectors and the VOD servers for transmission of the video data.

As part of the distributed infrastructure and application supporting features, the VOD service provisioning system also has some CORBA COSS services and interface objects including:

- *Trader*—a standard OMG CORBA trading service as described in the Chapter 5.

  The trader supports the increased fault tolerance of the system by locating alternative CORBA objects in the event of some system failure. For example, the session manager selector can locate alternative session manager factories via the trader. A session manager can also use trader to locate an alternative connection manager or a VOD manager if the existing one fails.
- *Supervisory console*—A management interface is provided on all CORBA objects. This interface allows the supervisor console to monitor the state of the CORBA objects.

8.2.3.4 Integration. There are three integration objects in this example system:

- Session manager selector
- Connection manager
- VOD manager

These objects form the integration points between the CORBA section and the CMIP section in the VOD provisioning system.

Let's look at the connection manager for more detail. The connection manager manages the network connections required for the VOD services. The connection manager does this on behalf of the session manager. At this integration point, a set of abstract CORBA objects representing connections are defined and it is based on, but not necessarily the exact mirror of, the real network, network elements, and network connection objects which are represented in GDMO. The connection manager has the knowledge of:

- How to map the abstract objects to the basic network objects
- How to map a service connection request to a set of network connection requests
- How to inform the session manager of which services might be affected in case of a network failure reported by the CMIP system

This knowledge is defined by the CORBA object modeler as part of the semantics (behavior) of the CORBA objects, and they form the implementation of the CORBA objects.

The reason to use the abstract object representation is that the connection definitions in the service management layer (VOD service definition) need to be modeled considerably differently from the standard GDMO definitions for connections. It needs to represent the end-to-end characteristics of the connection for the cross network technology domain purposes to greatly simplify the concept of the connection in VOD service provisioning. Also at service management level, some other service characteristics need to be incorporated, for instance, the quality of service requirements and the cost requirements. A one-to-one mapping would require the modeling of VOD connections to adopt the underlying object model. This will unnecessarily result in a complex object model in the VOD provisioning system. After this one-to-one mapping, the CORBA objects still require the level of abstraction to incorporate the high-level concepts discussed here.

The connection manager has two types of interfaces. One is the north-facing IDL interface to support the communication with the session manager, and the other is the south-facing CMIP interface to support the communication with the network and network elements. Through the IDL interface the connection manager provides an abstract view of network connections to the session manager which has no interests in network details. Through the CMIP interface, the connection manager manages network connections based on a complete view of the managed network and network elements.

Compared with the connection manager, the other two integration objects, the session manager selector and the VOD manager, represent much simpler cases. In the CORBA section of the VOD service provisioning system, the modeling of these objects may not differ from the underlying GDMO object definition. In this case, we could adopt the gateway approach which maps every CMIP objects into the CORBA space. However, even in this situation, we may still use the abstract object definition approach. Because in the VOD service provisioning system, we may want to extend the object interfaces to include more operations which are required by the VOD service provisioning system.

In the VOD service provisioning system, the objects managed by the session manager, such as service session, service user, are examples of service objects shown in Figure 8.11.

Objects managed by the connection manager, such as VOD connections, are examples of abstract network objects.

The case study presented in the Chapter 6 demonstrated how a generic network model can be designed to provide an abstract network view. This example further shows how the service and network integration can be achieved by the definition and implementation of this generic network model. These two aspects contribute different values to the CORBA/OSI integration. The combination of these two aspects makes a key component in our integrated management architecture. This will be discussed in the later chapters.

8.3 APPLICATION INTERWORKING

As stated before, one of the important purposes of the technology integration is to achieve the application interworking. More specifically, we would like to address the issue of executing business transactions involving multiple applications. Each application has been developed using different technology and with different object models.

The one-to-one CORBA/CMIP mapping approach introduced in the Section 8.1 primarily deals with a lower level (EML and NML) integration where specific technology mapping adapters are used to mediate different subsystems. The primary goal and the value of the approach is the protocol translation between two specific technologies.

The abstract mapping approach presented in Section 8.2 describes a process of a more coherent and technology independent object modeling and the implementation of different portions of the object model using different technologies to achieve integration. The value-add of this approach is manifested by a process of object abstraction, or a semantics enrichment, through the integration, not just the protocol translation. This approach primarily deals with the scenario of vertical integrations.

The scenario of application interworking is slightly different. It addresses the horizontal integration of different operation systems with different technologies and different object models. A typical example of this integration is the interworking between different service providers' operation support systems.

It is neither necessary nor possible for every service provider to accept the same technology and same object model and process model to implement their respective OSS systems. It is, however, crucial for different service providers to exchange application information between their OSS systems for the support of integrated, one-stop-shopping customer requests. In the new deregulated environment, this is important even just for the purpose and to the extent of fulfilling regulatory obligations (for example, the FCC interconnection requirements).

At a glance, this type of integration goes beyond the normal solution development life-cycle management discussed in this book. However, how one's operation systems behave to the external influences and interactions is every system designer's responsibility, and it has to be governed by the same coherent management strategy.

The application interworking typically requires a common interface between two operation systems to be defined and agreed. This interface agreement provides the basis for the application integration. However, this interface agreement has to have the following features:

- *It needs to be technology neutral*—As the target operation systems are likely to be developed using different technologies and by different suppliers at different stages, it is reasonable to suggest that the technology neutral approach to this interface specification has a chance to receive more acceptance. This technology neutrality does go beyond the notation used in the modeling process. It includes and focuses on the features and functions of the modeling languages and frameworks. For instance, a GDMO modeling technique implies a specific naming convention, and an object invocation through management entity, etc.; whereas in CORBA, the functions in all COSS services imply that they are supplied as external services in the computational architecture. The technology neutral approach should not assume the use of these implementation specific features.

- *It needs to be generic*—The interface specification has to be generic to fully encapsulate the specific design features and functions of different OSS systems. It should only cover the most general functions two systems are required to interwork. This also helps reduce the migration cost for the integrating operation systems. As a general rule, a subsystem supporting a very general interface is much easier to be replaced than a subsystem supporting a very specific interface.
- *It needs to be simple*—The interface specification should be defined as a minimum set of specifications only to exchange information for interworking purposes. It does not have to be an operational system in its own right.

The simplicity criteria of the interface specification is crucial to the success of achieving interworking. Many standard specifications, including many ITU-T and OSI standards, were originally designed for the purpose of interworking and interconnection (so the name OSI suggests). In reality they looked more like specifications of the operational systems. As such they became unnecessarily complicated. The paradox is that they are too detailed for the interoperating purposes (want to specify everything), but not enough for the operational purposes (cannot specify everything). This has resulted in the fact that the conformance of these specifications is very hard to define and enforce.

In the remainder of this section, we use the integration of service ordering systems between different service providers as an example to illustrate the issues and approaches of the application interworking.

8.3.1 Application Interworking Scenarios

When a customer requires a new service, the customer contacts a service provider. In many cases, the provisioning of the required service involves more than one service provider. For example, a request typically involves local telephone carriers and long distance carriers. From the customer's point of view, it is convenient that the customer deals with only one service provider (main service provider) while the main service provider orders components from the other service providers if required. This main service provider is the unique customer contact point with regard to the service request. In doing so, it is responsible for all communications with the subcontracted service providers for the purpose of exchanging information including service enquiries, service orders, billing enquiries, problem reporting, etc. A service interface model is such a model to support this level of interactions between two service providers. This means that the interface model has to interact with the service ordering operation systems at both ends of the interaction. The same interface model in this scenario also needs to interact with other parts of OSS systems (billing, provisioning, trouble reporting, etc.) within the main service provider, or with the peer service providers.

This leads to a need for a service interaction model to be used for application integration within an SP environment and SP-to-SP interworking. This requirement will be discussed in the next chapters. In this section, we focus on the service ordering interface model as the example.

It is quite conceivable that this same interface model can be used to support the interactions between customers and their main service providers. Figure 8.15 depicts the scenario of service ordering interface model being used for customer-to-SP interactions and SP-to-SP interactions.

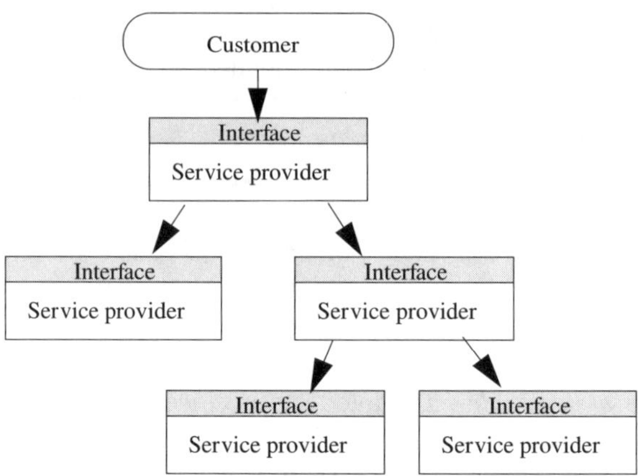

Figure 8.15 Service order interactions.

8.3.2 SP-to-SP Service Ordering Specification

The SP-to-SP service ordering is concerned with the ordering of services between telco service providers. In all service ordering activities, the requesting service provider can be viewed as a customer of other service providers with regard to the services provided in a typical producer/consumer fashion.

The specification deals with two issues: the agreement on the ordering business processes and the agreement on the ordering information model (or interfaces). There are two major phases in a service ordering process: *pre-order* phase and *ordering* phase. Interface objects are defined for both phases. The interface operations for the pre-ordering phase provide operations for a customer to make pre-order requests for services. The interface operations for the ordering phase provide operations for a customer to make service orders. The interface can also be used by the customer to query the status of both pre-order requests and service ordering requests.

The interface specification for the SP-to-SP service ordering focuses on the interactions between different service providers for service ordering purposes. How an ordering request is processed internally is not made visible at the interface level—this is the internal behavior of the ordering OSS. This follows the rules discussed before that the interface specification needs to be generic, simple, and technology neutral. More importantly, this interface specification only specifies the intended behavior and information passed between two interworking systems, not the internal behaviors of the two subsystems. Each service provider may have its own business rules. The handling of an ordering request should conform to these rules. Also, there are existing systems used by different service providers which provide basic capabilities of handling service ordering requests. It is neither practical nor necessary to require all existing systems to conform with a standardized internal behavior.

8.4 EVALUATION OF APPROACHES

Each of these three integration approaches discussed in this chapter has its own characteristics and application areas. The benefits of the IDL-GDMO gateway are manifested in the following ways:

- The concept of this gateway approach is easily acceptable. Industry's initial criteria for acceptance of a new technology or a new protocol is that it will provide at least the same functionality as the existing ones. To measure that, people will naturally try to define a close mapping between these two technologies. This human reaction has resulted in the close mapping approach between CORBA and OSI to be the initial major focus of many industry groups. This *whatever you can do, I can do* mentality is in itself a value-add that it made possible for multiple technology coexistence.

- There is an immediate high demand for interoperability between different network management components developed using CORBA technology and OSI technology. There is a real need for a CORBA manager to work with an OSI agent, or vice versa. With the support of this tool, it becomes possible for the managers and agents to be developed separately using different technologies.
- With the work conducted by consortia such as TMF, X/Open, and OMG, standard mapping algorithms are starting to mature. This will assist the development and deployment of this mapping approach. The solution of such a mapping is reasonably clear.
- The commercial environments for this mapping exist today. These include various commercial ORBs and network management platforms that support OSI communication.
- The perceived benefit of JIDM approach is that it provides tools for CORBA developers to get IDL definition for all currently defined managed objects. There is a saving on the object modeling exercise.

Potential problems also remain when this approach is used:

- Since the mapping between GDMO objects and IDL objects is a purely one-to-one mechanical translation, it does not add more value to the existing network management world, other than the same functionality that can be supplied by another technology. In fact, this approach loses some advantages of both CORBA and CMIP. For example, some flexible object management capability and the object encapsulation feature in CORBA are compromised to suit the CMIP and GDMO technology features.
- The approach does not solve the issue of producing a proper CORBA information model for network management and service management functions, thus to achieve the information abstraction and reduction. The mapping of GDMO objects and OSI functions has in effect polluted the CORBA object model with the unnecessary complexity of the GDMO object model and particular OSI technology features. Programmers are using CORBA to implement essentially an OSI application framework with all the OSI concepts (scope, filter, management entities, containment tree, etc.). To a certain extent, it is difficult to justify what value CORBA technology adds if it is only used to implement OSI technology features.
- In order to build more semantics into CORBA objects, particularly at the level where objects are representing management application entities and services, the environment to achieve object abstraction is still required and it is not provided by the tool. That is, the task of build higher level object abstraction from the existing object models cannot be achieved by the mapping.

The abstract object definition approach as described in this chapter shares the same value of a standard object-oriented process and methodology. It manifests the paramount importance of a *properly* defined object model for the business problems in an implementation technology neutral way. That is, the business problem modeling needs to be performed *before* any implementation or integration strategies are considered. In the scenario of CORBA integrated downwards with CMIP, the approach simply states that the object model the CORBA subsystem uses is not going to be dictated by the underlying CMIP system. It will be purely derived from the business requirements, and these requirements exist *before* the technology integration requirements; whereas the resulting CORBA object model from the one-to-one mapping is determined by the original GDMO object model and the translation algorithm. In many cases this trade-off for an automatic translation is not worthwhile.

To support the notion of the right technology for the right domain, the abstract mapping approach offers a better use of CORBA and OSI technologies. Given the major differences between the OSI CMIP architecture and CORBA architecture, it is a very inefficient use of CORBA to create a complete OSI CMIP model.

The application integration approach deals with a different class of problems. It requires an application interaction model to abstract the application semantics of two interworking systems. Therefore, the abstract mapping approach offers better support for application interworking. The principle of abstract mapping approach can be used to define this interaction model and to implement the interworking gateways based on the model.

To summarize the main features of three types of integration approaches discussed in this chapter, we use the following three statements:

- Whatever you can do, I can do (by closely mapping functions).
- Whatever you can do, I can do better by adding more value (by abstraction).
- We may not be able to, or need to, integrate completely; we can interwork to a degree that a minimum (and essential) set of requirements are satisfied.

All three approaches can be used in an integrated architecture to satisfy these three types of requirements. The IDL-GDMO gateway approach is more suitable for the interoperability between network management systems. It offers limited value in service management and network management integration. The abstract object definition approach is more suitable for the service and network management layers integration. It is also capable of supporting integrations between different network management technologies, such as CMIP and SNMP, by using the same abstract object definitions at CORBA layer. However, its value as protocol interoperability technique to CMIP and SNMP is limited. The application interface approach is more suitable to support the interworking between different service providers at the service management layer. It is not suitable for network level interoperability or network to service integration. Figure 8.16 depicts an integrated management scenario where different types of integration approaches can be used for satisfying different requirements.

A critical technology issue in an integrated telecommunication management architecture is the definition of an integrated information model. This information model defines the

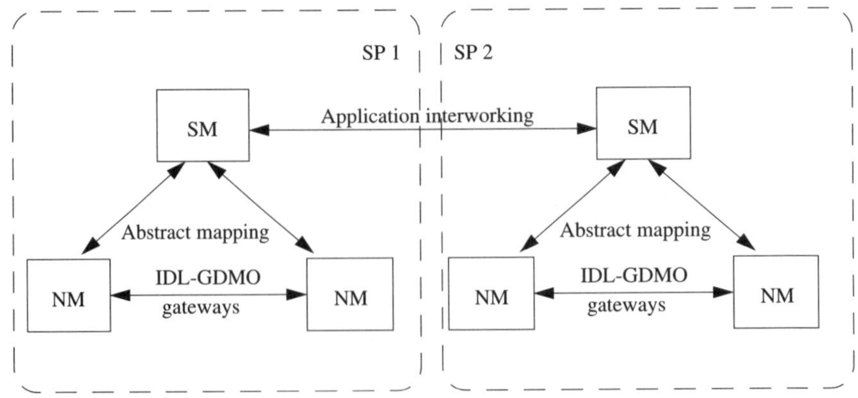

Figure 8.16 Applications of different integration approaches.

management activities and information flow-through from customer interactions with the management system, through to all different layers of the TMN architecture and across all different management functions. Only once this information model and process flow-through are properly defined, can the true integrated architectures and management solutions be achieved. This information model and process flow-through is at the heart of the management of the business process life cycle, as it provides us with a description of the business purposes and activities the integrated architectures and software components are meant to implement.

In the next two chapters, we focus our discussions on the definitions of these information models and process models. We define a TMN service interaction model and a generic network model to act as the frameworks for integrating different OSS and network management functions. We then present a software engineering solution architecture to implement these models using the approaches discussed in this chapter.

9

Generic Object Model

When we discuss the processes, information flow-through, and the integration between different TMN layers, one of the most important issues is the integration of the different object models. In TMN framework, these object models are the descriptions of the components of the TMN systems. These include the descriptions of the network elements, networks, and services. These models, called the *Management Information Base* (*MIB*), provide the basis for TMN management functions.

However, the object models so far were designed to reflect the specific details of the networks and their elements; they are often very complex. While the complex descriptions of the networks and elements are useful for the purpose of control and management of the equipment, it is hardly useful and necessary to expose all these network details to the service management functions. On the other hand, there is a lack of well-defined and accepted object specifications for the functions and activities in the TMN service and business management layers. This imbalance between excessively detailed and complex specifications on the network side and too little specification on the service and business side is often the source of difficulty for the integrated TMN management solutions.

In this chapter, we address the issues of the object specification. It is not the intention of this book to provide a complete object model for integrated TMN service and network management. The definition of these models is the subject of many collaborated efforts of the industry at large. Our focus in this chapter is to demonstrate the principles and processes leading to the specification of the generic object models, particularly in the context of software engineering using distributed and object oriented approaches.

9.1 APPLICATION INTEGRATION

It has been widely recognized that in the existing TMN model, the gap between the service management domain and network management domain has not been bridged properly. This has always made difficult the seamless integration between managing the customer environment and managing the network environment. The reality is always that the two environments have very different views of the management world and these two views of the world are too hard to bridge. In order to provide this seamless integration within a single service provider's domain, two essential functions must exist: the support for network protocol interoperability, and the support for integration of service management functions and processes.

The reality is that these two functions have raised conflicting requirements for designers of network models. The support for network protocol interoperability requires a very detailed and fully specified network model, whereas the service management functions are better supported with a more generic and abstracted model.

The current ITU-T-based network models are primarily designed to support network management functions and protocol interoperability. Therefore, a new model is required to support service management functions and application integrations.

In addition to the network protocol interoperability and service management functional integration within a single service provider, the interworking between different service providers adds extra value in providing wider range of services to end customers.

In the past two decades, protocol level interoperability has been perceived as the major technical challenge to the telco industry and has been the major subject of the open standards movement. However, the main issue facing the telco industry in a deregulated environment, in the software development sense, is service and application integration. A telco service provider may be able to exchange information with a peer service provider at the bits and bytes level. The irony is that this very SP may not be able to exchange meaningful information between multiple service offerings and different management applications in its own jurisdiction.

The TeleManagement Forum has a whole SmartTMN program devoted to this purpose: to achieve a better flow-through of information, processes, and management functions between the service/application and network boundary. All these efforts point to the need for a more generic, abstract, protocol neutral and service management-oriented (north facing) view for modeling networks and network management functions.

Also, as mainstream distributed object technology such as CORBA is demonstrating its value to the telco industry by offering an application development and integration environment, the work in the CORBA/OSI integration area also provides input to work in generic network modeling.

However, there are two misconceptions in these exercises. First, some believe that using more protocol neutral notations to describe the network, such as OMT or UML, will serve our purpose. One approach is to reverse-engineer existing ITU-T GDMO specifications (such as M.3100) back to OMT or UML notation, and to use the result as the generic model.

Unfortunately, this approach fails to realize that the primary purpose of network modeling is to provide a better application environment to support service management functions. The end result of this reverse engineering, no matter what modern object-oriented notation is used, presents a very network- and element-management oriented (south facing) view. This view is intrinsically part of the original GDMO object definitions. The process is only an object model translation, not an abstraction. With the exception that the end result is presented in a different language, the object model inherits all the features, properties, and drawbacks of the original object model, without adding value.

Second, some CORBA/OSI integration research is also focusing on protocol translation. This approach treats CORBA as merely another protocol, along with CMIP and SNMP. However, a large part of the industry has realized that although CORBA can be used as a new protocol to support network management, it doesn't add much value over the current OSI technology if it is only used as a new protocol. Its main attraction is that it offers a much better application development and integration environment for TMN applications and service management functions.

The main problem stems from the inherited complexity of the network and element view of the network structure. When this view was developed, its very purpose was to provide a description of the telecommunication networks and their elements so that the networks and their elements can be configured and managed. Its main focus was not on how this view is used to support service and customer management layer functions. The complexity of the network and elements is represented by the complexity of the network model, the MIB. There is a vast quantity of information being modeled in the standard MIBs which represent every aspect and every detail of the network and elements.

Much of this information is required by the network operators and administrators to understand how the network is operating. However, for the service operators who manage services and customers, there is hardly any need to know this complex network information model. A service application responsible for provisioning a video-on-demand connection is only interested in an end-to-end connection between two end points with a certain quality of service features. This service application is not interested in the fact that this connection goes from an ADSL network to the ATM and SDH network and the connection is made up of different cross connections, subnetwork connections and link connections.

From the software engineering viewpoint, the complex object model increases the complexity of the management applications. In a typically small SDH subnetwork with two ring structures, one could instantiate tens of thousands object instances following the I-ETS 300 653 managed object class library.

What is required by the telcos, for the integration within one service provider or between different service providers, is a set of generic object models. The purpose of the generic object models is not only a protocol neutral description of the networks and services. A generic object model (GOM) has very little to do with protocols. It is designed also for the purpose of information abstraction, and it provides a basis for application integration. It supports an abstract view of the network services and/or a generic view of customer services. It tries to describe an SP's view of the network and helps an SP to map its business operations to the existing network functions.

While the GOM represents the generic network functions from an SP application's viewpoint, it is implemented, through object specialization, by different network models and network management functions. The models at different levels deal with different issues. The GOM maintains support for an SP's business application development and integration; it is independent from the underlying network models and network management functions; and it has the same life cycle as the business applications it supports.

On the other hand, the existing network models are defined to represent the network view from a network provider's viewpoint. They provide a full specification of the network functions, whether they are visible to applications or not. The life cycle of these models is related to the life cycle of the network technologies, and they are the points of protocol interoperability at the network layer.

In this chapter, we discuss the TMN service interaction model and generic network model. The former focuses on the interactions between different service management functions, processes, and service management operation systems. The latter focuses on the support for integrations between service management functions and the network management functions, and between existing network management technologies.

Considerable effort has been made by various standards bodies and forums to define the generic object models. These efforts have resulted in object models at different levels of TMN management hierarchy. In this chapter, we position the discussion mainly on the processes of generic object modeling with a strong focus on the development of software components for achieving application integration.

9.2 SERVICE INTERACTION MODEL

In this section, we focus on the definition of TMN service interaction models. A service interaction model defines the interaction interfaces between different service management activities and functions, and between different service management domains. It represents the external behavior of the service management functions.

From the software engineering and solution development's point of view, this service interaction model plays a pivotal role in supporting smooth operations of any service provider.

First, this model provides an upwards integration with the customer management environment. This service interaction model defines all services and service management activities to support customer's service interaction. The service interaction model is a manifestation of these activities for the purpose of interacting with customer care and management.

Second, this model provides a horizontal integration with other service providers' management environment. It is a basis for the information and functional exchange for the purpose of supporting cross-SP service management and wholesale–retail interface.

Third, this model provides the basis for the application integration within a service provider's environment. The service interaction model provides the application developers with a road map and this road map reflects the fundamental business requirements and is independent from technology choices for each individual software development exercise. It gives software developers a better chance to build integrated software solutions and to migrate them to newer technologies.

Although it is well recognized that lack of a well-defined business process model and information model is the main reason to prevent an SP to become a smooth operator, often the lack of standards is used as an excuse for individual service providers not to act upon. In fact, from service provider's viewpoint, it is far more important to have a well-defined business process and information model than waiting to have a standard one (if it is at all possible to have a standard one). Lack of this service model results in a typical scenario in which an SP can exchange usage data between their switches and a peer SP's switches using OSI protocol, but they cannot provide an integrated bill among their own service offerings.

A different integration requirement not addressed by this model is how to provide a downwards integration strategy with the network management systems. Although the customer service requests are finally supported by the network resources, there has to be an abstraction of the network functions and the representation of relationships between the network functions and the customer service requests. In the later sections, we present the concept of generic network model to represent the abstraction of the network resources and functions. We then discuss the integration between the service interaction model and the generic network model to achieve this downwards integration strategy with the network management layer.

9.2.1 Overview of the Service Interaction Model

The service interaction model presents an abstract TMN service view provided by an SP to its customers. A customer can be either an end customer in a customer–SP relationship or another service provider in an SP–to–SP relationship. For an end customer, the service interaction model supports various service management access points to the customer access applications, such as electronic customer access and customer service management systems. For other service providers, the service interaction model supports various integration points to the interworking system applications between different service providers. In both cases, the service interaction model is the access interface to a service provider's management environment.

An integrated TMN service interaction model should have the following characteristics:

- The model should be generic enough to be commonly applicable to most service providers and to support most common and essential functions. Any service specific features should not be included as part of the model. For example, the same service model should be useful to both local service providers and long distance service providers.

- The model should contain only intended system behavior manifested as the external interface specification rather than internal processes. It is important to let the development of service management applications define or specialize the behavior of the internal process. This is due to the following reasons:
  —Each service management system supports its own business rules. The handling of service requests should primarily conform to these rules.
  —There are existing service management systems used by different service providers which provide certain service management capabilities. It is neither practical nor necessary to require all the existing systems to conform to a standardized internal behavior.
  —As a design principle, the service model should mainly be used as a framework for interworking between different service providers and applications, as such it should not overspecify internal system behaviors.

A service interaction model can be partitioned into a number of functional areas to model different service management activities and processes. Each functional area can be defined as a separate interface model. These models contain the minimum object definitions to describe the breakdown of business processes and activities in specific functional areas. These models also contain the interface agreements for the purpose of interaction with peer service management functions. Each interface model can be implemented individually by a service provider or several interface models can be implemented to provide a more complete service management environment. Examples of such interface models are:

- a service ordering interface model, which supports all the service order related requests from end customers or other service providers.
- a performance interface model, which handles all the performance related issues such as quality of services, service level agreement, etc.
- a billing interface model, which handles billing enquiries, on-line payment, etc.
- a problem report interface model, which handles all the problems reported by customers.

Each interface model is supported by a management subsystem and this subsystem handles the requests issued through this interface. It is often required that these different subsystems work together to provide an integrated management environment. For example, during service provisioning, a service level agreement is normally reached between the service provider and the service user through the ordering process. The performance management subsystem should form part of the management of this service level agreement in order to achieve the agreed quality of services. The problem handling subsystem should also be part of the service level agreement management to support customer trouble ticket management and to provide speedy resolution of the reported problems. This is true also for the billing subsystem which calculates service charges and presents bills according to the actual service performance against the service level agreement.

In the remaining part of this section, we examine with more detail the interaction models. We address the following business processes and activities:

- Service ordering
- Problem reporting and handling

- SLA management
- Performance management
- Account enquiry and payment

The service interaction model is focused on the interactions between multiple business processes and the corresponding service management components. It is not focused on the operational nature of the processes. We will not discuss the internal operational behaviors and structures of each subsystem as the focus of the book is on the process and technology integration.

9.2.2 Service Management Activities

9.2.2.1 Service Ordering. Service ordering business activity deals with two scenarios. End customers can require the service and products offered by the service provider they choose to use. This interaction can be entered directly, or it can be entered after the sales process. During the sales process, customers may use the product and service catalog or other sales approach an SP offers to locate and select the products and services they are interested in ordering. The customers may order a product or service that is found from the catalog search with matching service descriptions, features, quality of service parameters, cost, etc.

Alternatively, customers may enter the so-called pre-order activity. During the pre-order phase, the customer can negotiate with the service provider for some specific service features and QoS parameters. Some of these services cannot be ordered as per catalog services, as they may require off-line design (not necessarily fully automated), combination, and configuration of some existing service offers. As such, the ordering status tracking is required. So is some simple management functions such as cancel the pre-order, and select from different service options.

Whether customers enter ordering system for a catalog service or a special designed service, a service level agreement is implied or specifically reached. This service level agreement includes the agreed service features, quality of service parameters, problem resolution agreement, cost and payment terms, etc. The service level agreement is used as an integration point between ordering process and other service management processes.

The ordering process can also be entered by other service providers who require components as part of their service offerings to their end customers. In this case, there should already exist a governing business framework (such as retail–wholesale agreement) which stipulates the overall service level agreements. Each individual instance of ordering activity can derive a specific service level agreement from it.

In the case of the end-customer initiated service ordering, the process supports the integration between the existing ordering operation systems and the customer care and access systems. In the case of the service provider initiated ordering, it provides an integration framework to interconnect two specific ordering systems existing in two service providers OSS environment.

9.2.2.2 Problem Reporting and Handling. Problem report and handling is concerned with the problems or difficulties experienced by the customers, whether the customers are end customers or service providers interacting on behalf of their customers. The problem reporting process interacts with the customer care and access applications as it forms an integral part of the customer management system.

When a problem is detected by customers who are using the service, the process must provide functions for a trouble report to be generated and sent to the service provider. A speedy response is expected by the party who issued the report.

On receiving a trouble report, the service provider interacts with its internal fault management subsystem and performance management subsystem to identify the cause of the trouble and resolve the problem. The process interacts with its internal service management and/or network management functions to locate and rectify the problem. In the event of two interworking service providers, the problem management process will interact with the service provider who is responsible for the service quality on behalf of its customers.

In the case of quality of service degradation, the billing subsystem, through SLA management, may be involved to reduce the service charges. This largely depends on the terms of the service level agreement when the service is ordered.

9.2.2.3 SLA Management. SLA management process is concerned with the overall management of the service level agreements. It coordinates different service management processes to achieve the overall customer satisfaction in terms of service quality.

SLA management interacts with the customer care and access processes by providing the status of the SLA monitoring and, in more proactive customer management systems, the forecasting of the quality of service trend and the warning of potential quality degradation. It provides the continuity in the life-cycle management of the customer interactions following the service ordering process.

SLA management achieves these by interacting with all service management systems such as fault and performance management, ordering, and accounting systems. It is a major horizontal integration point within the service management environment.

9.2.2.4 Account Enquiry and Payment. The account and billing process is concerned with the account management activities between customers and the service providers or between two service providers. The purpose of the process is to generate bills which accurately reflect the service usage, tariff rules, and the execution of the service level agreements.

The process integrates with the customer care and access system by supporting the customer access to billing and service charge records. Also it supports the bill presentment and payment functions. The process forms an integral part of the SLA management. It also interacts with the internal OSS system for creation, management, and storage of, and access to the resource usage records.

9.2.2.5 Performance Management. The performance manager is concerned with the performance monitoring of customer-ordered services. Service performance data is a service management level view of network performance data.

The performance manager supports interactions with other service management processes for changing or obtaining performance data. For example, it supports the interactions with the service ordering process for service creation and deletion information. It also interacts with the SLA manager to obtain SLA information in order to detect SLA violation.

Potentially, a performance manager can detect a service failure by examining service performance data and issuing a service fault report to the service fault manager.

9.2.3 Basic Interaction Model

The service interaction model describes service management by identifying the interface objects and their relationships to each other. These interface objects support all interactions required between service providers for the purpose of service management. They also support the interactions with the customer service management.

The following object classes are defined for service interaction.

- *A Service Provider (SP) class*—The SP class represents an organization which provides telco services. It supports operations for making a pre-order request, issuing a service order, etc.
- *A Customer class*—The customer class represents an organization which requires telco services from an SP. This class contains all information about the customer. It also contains references to a set of services the customer has ordered from the service provider and related account information. When a customer requests a service from a service provider for the first time, an object is created for the customer.
- *A Service class*—The service class represents a subscribed service provided by the service provider to its customers. A customer may order more than one service from a service provider, and a service provider may provide more than one service to a customer.
- *A Service Level Agreement (SLA) class*—The SLA class represents the service level agreement between a service provider and its customers with regard to a particular service provided. It includes the objectives (e.g., performance objectives) and other parameters regarding the provision of a service and the quality of services.
- *An Account class*—The account class represents an account created by the service provider for billing purposes. Between a service provider and a customer, a number of account objects can be maintained. More than one service for the same customer can be charged to the same account.
- *A Service Request class*—The service request class represents a pre-order request issued by a customer to a service provider. The customer can issue a query requesting the current status of the pre-order request. At any stage the customer can cancel the pre-order request.
- *A Service Offer class*—The service offer class represents a service offer made by a service provider to a customer in response to the customer's pre-order request. For each service pre-ordered by a customer, only one service offer is created. However, a service offer may contain a list of options detailing how the service can be provided in different configurations. These options are used by the customer as the basis for the service offer negotiation.
- *A Service Order class*—The service order class represents a service order request made by a customer. The valid requests that can be issued to the service order object are: to update the order, to cancel the order, and to query the progress status of the order request. This object does not represent a service instance. A service object instance is created when a service order is delivered and activated.
- *A Trouble Report class*—The trouble report class represents a problem reported by a customer to a service provided.
- *A Performance class*—The performance class represents the performance data associated with a service. It keeps all the performance related information of the service from the time the service is activated to the time the service is terminated.

The service interaction model is depicted in Figure 9.1. The relationships among the service object classes are:

- Each service provider is represented by a single SP object
- Each customer is represented by a customer object
- Each customer can have one or more accounts with a service provider. Each account is represented by an account object. More than one service can be charged to a single account.

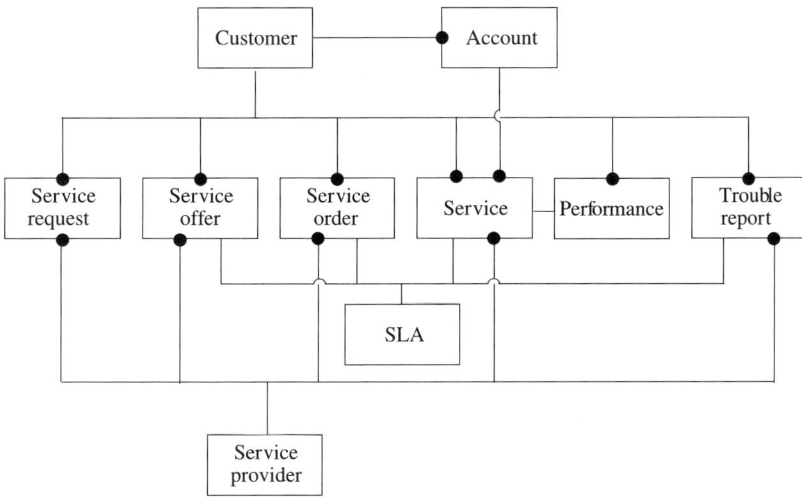

Figure 9.1 Object instance model for service interaction.

- Each customer can have more than one service request instance associated with it. Each service request represents a pre-order request made by the customer.
- A service offer is made by a service provider to a customer that satisfies the conditions of the customer request and the negotiated service parameters. A service offer can be made for each service request and is represented by a service offer object.
- Each customer can have more than one service offer instance made by a service provider.
- A service order is made by a customer. Each service order is represented by a service order object. Each customer can have more than one service order instance associated with the customer request.
- A service level agreement is made between a customer and a service provider with respect to a service. Each service level agreement is represented by an SLA object. During the pre-ordering and the ordering phases, an SLA object instance can be associated with a service offer or a service order object instance, respectively. After a service is ordered, the SLA instance is then associated with the ordered service.
- Each service is represented by a service object. Each service object has an SAL associated with it. Each service will be charged according to the account associated with the service. Different services may be charged to a single account.
- The performance of a service is represented by a performance object. Each performance object is associated with a service object.
- A trouble report can be made by a customer with regard to a single service or a number of services provided by the same service provider. Each trouble report is represented by a trouble report object instance. The report may refer to the SLA object instances associated with the services reported.
- A service provider can support multiple service requests, service offers, service orders, and services for different customers.
- A service provider can handle multiple trouble reports issued by different customers with regard to different services.

A number of service provisioning and management scenarios are presented in the next subsection to demonstrate how the interaction model can be used.

9.2.4 Service Provisioning and Management Scenarios

An important part of the service interaction model is the interaction behaviors between different processes and interface objects. This subsection describes these interactions by using different interaction scenarios.

9.2.4.1 Service Pre-ordering Scenario. A pre-order request is issued by a customer to a service provider. A customer uses the pre-order request to specify the service requirements including its features, characteristics, and quality of service requirements. The service providers use the pre-order phase to supply offers (perhaps with multiple options) to the customer, and the customer uses pre-order phase to select, compare and negotiate an offer.

The process begins with the customer issuing a pre-order request, which may follow the process of searching the product and service catalog as part of the sales process. The pre-order process can be completed in a number of different ways. A successful pre-order scenario is described as:

1. A customer issues a pre-order request to a service provider for a certain type of service.
2. Upon receiving the pre-order request, the service provider creates a service request object instance for the pre-order request and sends a response back to the customer with the service request identifier. This identifier can be used by the customer for status tracking and request cancellation.
3. The service provider decides to make a service offer, and it sends a reply to the customer with some details of the offer including the initial SLA associated with the proposed service.
4. The customer and service provider may enter into further negotiations to finalize the choice of the options and the service features.
5. The customer is satisfied with the negotiation result and is willing to proceed by making a service order within the valid time frame specified in the offer. The pre-order request is regarded as complete when a service order is placed by the customer.

At any stage after the pre-order request is made and before a service offer is received, the customer may query the progress of the request. The service provider should respond the queries by issuing the status of the request.

At any stage after a service offer is made, the customer may query the progress of the negotiation by requesting the status of the service offer object. The service provider should respond to the queries by issuing the status of the negotiation.

At any stage after the pre-order request is issued and before the service is ordered, the customer can cancel the pre-order request. This results in an unsuccessful termination of the process.

At any stage after an offer is made and before the service is ordered by the customer, the service provider can cancel the offer. This also results in an unsuccessful termination of the process.

9.2.4.2 Service Ordering Scenario. A service ordering request is made by a customer to a service provider to place an order for the identified service. The service ordering can either be based on the result of a pre-order request or on information obtained from other sources such as the catalog search.

All service order scenarios begin with a customer service order request. A successful service ordering scenario can be given as:

1. The customer issues an order request.
2. Upon receiving the order request, the service provider creates a service order object and starts its internal ordering processes. It then sends a reply back to the customer with the reference to the order object and the expected processing time for the order. The operation of the service ordering processes is internal to each service provider and not part of the service interaction model. It may include some of the following activities:
   —develop an order plan
   —configure the network resource required for the service
   —place suborder(s) if required
   —inform other processes, such as billing, performance management about the service order
   —create a service object when the order has been implemented
   —perform internal service testing before informing the customer
   —create/update SLA through SLA management process
3. Once the order process is completed, the service provider informs the customer of the availability of the service with a reference to the service object and the finalized SLA object.

At any stage during the order processing, both the customer and the service provider can cancel the service order. This results in an unsuccessful termination of the ordering process.

9.2.4.3 Problem Reporting and Handling Scenario. A problem report scenario is triggered by a customer reporting a problem experienced when using an ordered service. In some cases, a trouble report may involve a number of services provided by the same service provider. A response is always expected from the service provider unless it is cancelled by the customer. The response is to explain the reasons for the problem, whether it is a service fault, and, if so, how it was resolved, etc.

A common problem report scenario can be described as follows:

1. A customer issues a problem report to the service provider with reference to the services affected by the problem.
2. A trouble report object instance is created by the service provider for each problem reported.
3. A sequence of internal problem report handling activities is invoked. These activities interact with internal fault management system and performance management system to identify and locate the possible cause of the problem. It may be required to communicate with the customer for more detailed supplement information.
4. If a fault is identified, some internal processes may be invoked to repair the fault.
5. Depending on the characteristics of the problem and the service level agreement, it may be required to make billing adjustments as part of the overall SLA management.
6. The service provider needs to inform the customer after the problem is resolved and the billing adjustment is made.

9.2.4.4 SLA Reporting and Monitoring Scenario. SLA reporting and monitoring scenarios are usually a simple reporting process. Customers are provided with the SLA monitoring information which reports the current execution of the services with respect to the SLAs.

9.2.4.5 Account Enquiry Scenario. Customers can issue account enquiries at any time after services are provided. For each request, customers provide information for the service provider to identify the account, such as the account number or the combination of service ID and customer address.

An account enquiry scenario is described as follows:

1. A customer issues an account enquiry request to its service provider with the customer ID and the account number or alternative service identification.
2. An internal billing management system is invoked by the service provider to obtain the required information. The processing of such a request is internal to the service provider.
3. A response is sent back to the customer with the information required.

In many cases, the process requires the interactions with the SLA management. It enables the more accurate presentation of the accounting information to include, such as the compensation resulting from the service degradation.

9.2.4.6 Account Payment Scenario. A customer can activate a payment request through the customer service interface. Extra security management, such as encryption, may be required to support the request. This scenario can be described as:

1. A customer issues an account payment request with its customer ID, account number, and payment details, such as bank card number, expiring date, payment amount, etc.
2. The service provider, upon receiving such a request, may invoke internal payment processes to verify the customer, service and account details, and process the payment.
3. A confirmation and a reference number for the payment need to be sent back to the customer.

9.2.4.7 Performance Reporting and Monitoring Scenario. The performance reporting and monitoring scenario is usually a simple reporting process. Customers are provided with the performance monitoring information which reports the current service performance and with a reference to the SLA.

The service interaction model and the interaction scenarios define the behaviors of the service management processes in terms of their interactions. These processes support the customer care system for the customer management purposes, and the service management applications. They support the interactions with the peer operation systems for interconnection purposes.

Chapters 10 and 11 discuss the process model and software architecture that support the interaction model. They present more details on internal processes, their roles in the complete integrated management architecture, and software implementation issues such as technology selection and evaluation. The next few sections in this chapter present how this model is further extended to the network domain to provide integrations with the network management functions.

9.2.5 Summary

This section presents a process to define a service interaction model to support the application integration and interworking. For each individual service provider, there are obviously the needs to produce a more comprehensive service interaction model using more established object modeling techniques, which is outside the scope of this book. The key point is that this modeling is crucial to the success of the business process reengineering and to the success of supporting customer services. This chapter and the next two chapters demonstrate how these models can be combined in a solution architecture supporting integrated management.

It is important that this model be developed with the following views:

- It is developed to support the most essential business activities.
- It is mainly used as an integration device for different existing and new applications.
- It is the interaction point for supporting customer services.
- It defines the interworking access points for SP-to-SP interconnections.

In order to extend the useful life of the service interaction model, the model has to be generic and simple. We must remember that this model is not an operational model but an interaction model. This model, together with the existing service and network management applications and existing OSS systems, form the complete implementation of the service management system. We will never have an opportunity to completely remodel and redevelop our entire service management systems at any given time. We have to integrate with older systems and someone else's systems which may not fit into the model.

When the model is generic and simple, it extends its life cycle, and we have more room for future changes. These changes include the upgrade and replacement of our network management systems, OSS systems, and applications. They also include the underlying network or IT technology we use to manage networks and services. These changes are always going to happen, and they happen at a much faster speed than the changes of the way we do business. A generic and simple model helps us preserve the business rules by replacing components.

As stated before, this service interaction model is used to support more generic access from the customers to network functions. From the viewpoint of object orientation, the service models need to be mapped down to the underlying network model which supports services.

The underlying network functions are usually supported by different network technologies with different network models. On the one hand, these models were developed from a very strong viewpoint of network and element management. They usually contain very specific network details. On the other hand, an integrated service model requires a very high level abstract view of the network to enable the end-to-end management functions. For example, a service provisioning system typically would like to request: *find me a connection from A to B in the network (usually cross different technologies) with these service parameters and within this cost range.* Furthermore, for the benefit of extend the useful life of the service management system, the implementation of the service model should be independent from the specific details of the underlying network technologies and the management technologies used in network management layer. A more generic and simplified network model is required which represents a service management view of the network functions and strongly supports the end-to-end service management activities.

In the next section, we examine this generic network model and the issues to relate this to the existing TMN network models.

9.3 GENERIC NETWORK MODEL

In this section, we discuss the issue of TMN network modeling. We go through a process of defining a subset of a generic network model [20] as an example to demonstrate our approach. We use the modeling of network topology and connections as examples. In this process, we focus our attention on the following issues:

- The definition of the network topology model
- The definition of the network connection model
- The specialization of the models to the more detailed and technology specific network models

In the case study presented in Chapter 6, we discussed a more generic network model based on the ITU-T G.805 architecture and the ITU-T M.3100 network model. Our experience showed that a generic model is very useful in two aspects:

- Developing a management application with more end-to-end, cross-technology domain functions
- Developing a better integrated application with multiple management technologies (CORBA and OSI).

The experience also showed that although a management application needs a more detailed and technology-specific object model to be functional, it also requires a more generic object model to be able to integrate upwards with the service management applications.

The purpose of the generic network model is to provide an integration point between service management functions and applications to network management functions. The generic network model acts as the integration platform to support:

- The mapping between service management functions and network management functions supported by the ITU-T based network model
- The abstraction of different network topology and connection models (ITU-T M.3100, ITU-T G.805, ATMF M4, and ETSI 300 653, etc.).
- The abstraction of topology management and connection management functions offered by different network technologies (such as ATM, SDH/SONET, WDM and IP).

This generic network model is not designed for the purpose of protocol interoperability. Its multiple specializations into individual network technologies and models are supporting protocol interoperability as indicated in Figure 9.2. The generic network model supports application integration and provides a stable environment for life-cycle management of telco solutions in a multiple technology environment.

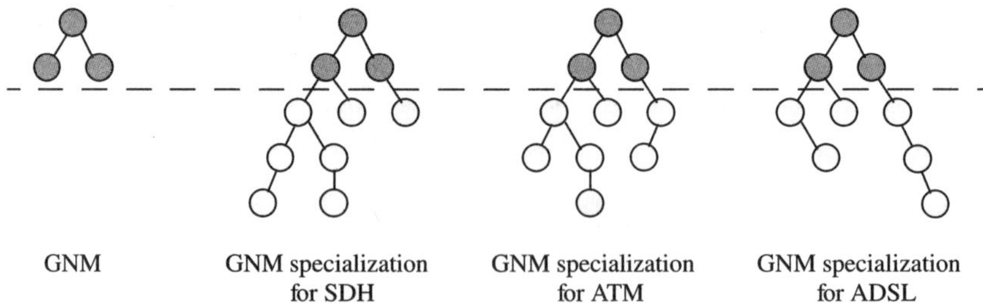

Figure 9.2 Generic network model and its specializations.

The generic network model contains the following components:

- A set of network objects which form the basis of the object model
- A set of operations supported by the network objects. These operations represent the service capability supported by the model, and they can be used by the service layer as the access points.

- The management functionality which includes TMN FCAPS functions.

With different management focuses, the generic network model can be partitioned into different submodels. For example, a fault management model may support the full functionality in the fault management area and limited functionality in other management areas; a connection model may focus on the network configuration functionality with limited support for other management functions, etc.

In this section, we focus on the principles and processes used to define the generic network model. We go through an exercise to define a topology submodel and connection submodel to illustrate these principles and processes.

The topology and connection models support service management functions where generic end-to-end connections are required. It is designed as an integration device between the network management environment and the service management environment. It is aimed at integrating with the service interaction model discussed in this chapter. The topology and connection models are not designed to provide protocol interoperability. The existing ITU-T connection models and any proprietary connection models are treated as specializations of the generic network model. This relationship is that of object generalization and specialization, not of object models translation.

9.3.1 Generic Network Topology Model

A generic network topology model is a subset of the generic network model. It is a framework for managing network topology supporting telecommunication systems. The network topology model contains the following basic components:

- A set of topology objects and the operations supported by these objects. These operations can be used by topology management applications to construct and manage network topologies.
- A set of semantic rules to be used to define the behavior of the network topology model. Examples of such rules are the containment relationship of network and its subnetworks, the visibility of subnetwork internal structure, etc.

The main concepts in the network topology model already exist in ITU-T's generic network architecture. However, the ITU-T generic network architecture is designed for the instrumentation of the network management systems. It is used for the purpose of:

- Modeling the network and element details
- Implementing network management functions and systems
- Facilitating protocol level interoperability

The generic network topology model described here is used as an integration device for the purpose of:

- Supporting service management functions in a network technology neutral way
- Providing abstraction for cross-domain network topology management purposes
- Supporting network and service management application integration

9.3.1.1 Basic Object Model. The topology model contains a number of topological objects. These objects are used to represent the logical and physical structure of the networks to be managed. There are three basic objects in the network topology model.

- `Network`—a network object is a representation of a collection of interconnected logical and physical management objects capable of exchanging information.

  A network may contain a set of subnetworks. The containing network is often referred to as the super network. Each subnetwork itself is a network and can contain some other subnetworks.
- `Termination`—a termination object is an access point of a network. Terminations can be combined to form termination groups. This concept allows the terminations to be viewed and manipulated at different abstraction levels. In the remaining section, we use the term *terminations* to refer to both terminations and termination groups. The network connectivity is managed through the use of terminations.
- `Link`—a link object represents a topological link between two subnetworks through two terminations, one in each subnetwork. A link belongs to the super network containing subnetworks and their links.

Network, termination, and link are closely related concepts. The relationship between these topological objects is depicted in Figure 9.3.

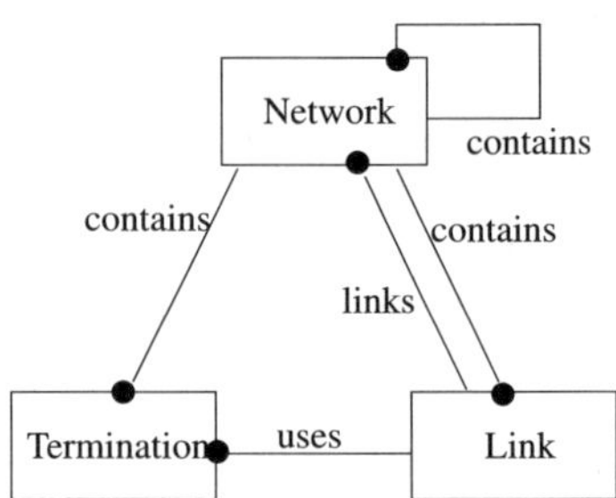

Figure 9.3 Object relationship in topology model.

Networks contain terminations. A network offers its services through terminations. A termination exists in only one network. No terminations can be shared between networks. Terminations in one network usually represent customer-visible network access point. Networks are connected by links through terminations. Both the connected networks and the link connecting them belong to the same super network.

9.3.1.2 Basic Network Topology Services. The topology model supports a set of basic network topology operations through the interfaces on its objects. The following operations are used to construct a network topology:

- *Create network*—The operation creates a network instance in a network topology.
- *Delete network*—The operation removes a network instance from a network topology.
- *Add subnetwork*—The operation adds a subnetwork instance to a given network topology.
- *Remove subnetwork*—The operation deletes a subnetwork from a given network topology.
- Add termination—The operation adds a termination to a network topology.
- *Remove termination*—The operation removes a termination from a network topology.
- *Create link*—The operation creates a link between two subnetworks contained in a super network. This link connects two subnetworks using their terminations.
- *Delete link*—The operation deletes a link between two subnetworks.

In addition to the basic construction operations, the topology model also supports a set of basic search operations for the retrieval of the network topology information. These search operations are:

- *Find termination*—The operation searches for all terminations in a given network topology.
- *Find link*—The operation finds all links contained within a network topology.
- *Find subnetwork*—The operation finds all the subnetworks in a given network topology.
- *Find link termination*—The operation returns both terminations of a given link.
- *Find termination link*—The operation finds the links using a given termination.
- *Find linked subnetwork*—The operation finds subnetworks linked by a given link.

More operations can be constructed using the above basic operations.

9.3.1.3 Internal and External Views. Each network has two views, an internal view and an external view. The internal view captures the internal characteristics of a network itself, for example,

- How many subnetworks are contained in this network
- How the subnetworks are linked
- How much information is made available by each subnetwork (i.e., the external view of each subnetwork as described below)
- How the information about the subnetwork's external view is used in this network, etc.

The external view represents the characteristics of a network made visible to the outside world; this view includes the network itself and a set of external terminations of this network. The concepts of internal view and external view are useful to define the relationships between a network and its subnetworks.

Figure 9.4 illustrates the internal and external view of a network containing two subnetworks.

9.3.1.4 Networks and Subnetworks. A subnetwork itself is a network. The network object is used to represent either a network instance or a subnetwork instance. The relationship between a network and its subnetworks is a simple containment relationship. A network

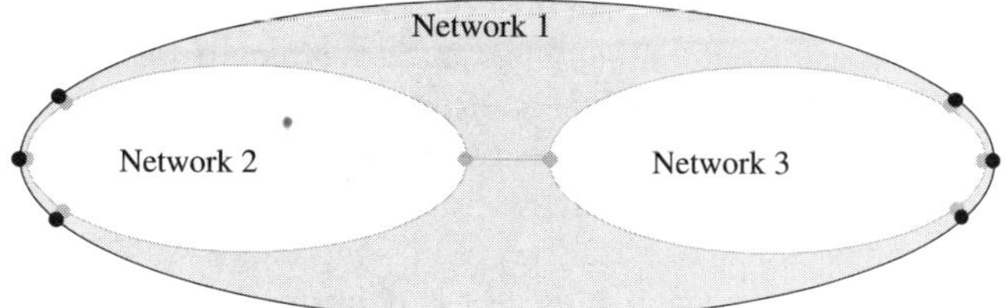

external view of a network and its terminations
subnetwork and its external terminations (as part of the internal view)
link between subnetworks (as part of the internal view)
direct mapping of subnetworks external terminations in the supernetwork

Figure 9.4 Network internal view and external view.

contains a number of subnetworks, and a subnetwork is directly contained only in one network. This containing network is referred to as the super network or a parent network with respect to its subnetworks.

A super network can have the external views for each of its subnetworks. It also has the knowledge of the relationships between its contained subnetworks. For example, a super network knows the external terminations exposed by its subnetworks. It also understands the connectivity (i.e., the links) among its subnetworks. However, it is not necessary for a super network to know the internal views of its subnetworks which reflect the internal topological structure. For example, whether a subnetwork contains its own subnetworks and how these subnetworks are linked is not visible to the super network. As a simple rule, a network only has the external view of its subnetworks and it only exposes its external view to its super network. This rule allows the super network to represent a higher level abstract view for its subnetworks.

The relationships between a network and its subnetworks can be summarized as:

- Each network is a logical representation of a group of subnetworks. It provides a layer of abstraction for these subnetworks. Externally, a network can be viewed as a single entity.
- Each subnetwork can publish a set of terminations as its external terminations. These external terminations are visible to its immediate super network but not to any other networks. If the subnetwork offers services, then these external terminations are the service access points.
- External terminations of a topmost super network are the access points for services provided by the entire network. They typically provide an end-to-end view of the entire network functions.
- An external termination of a subnetwork can be made visible at the super network level. This is done by defining an abstract termination in the super network as the direct mapping of the corresponding termination in the subnetwork.
- A super network can publish its external view via its external terminations. Abstract terminations can also be published as external terminations.
- The connectivity of subnetworks is represented in the super network as a link between subnetworks. Only the external terminations of the subnetworks can be used to create links. In other words, subnetworks can only be connected through their published external terminations.
- A subnetwork has no knowledge of how it is linked in its super network.

9.3.1.5 Hierarchical Structure of the Network. The network and subnetwork containment relationship determines a hierarchical structure of the entire network. Such a hierarchy structure can be illustrated by an example in Figure 9.5.

Figure 9.5 depicts two different views (an external view and an internal view) of a network. For a top-level network, the external view is used to publish its service access points to the service management applications requesting end-to-end functions. For a subnetwork, its external view is used to publish its access points to its super network.

For each network, the internal view contains topological information visible only to itself. It also contains the relationships between the current network and its subnetworks.

In Figure 9.5, `network1` represents the entire network. It has six external terminations through which services are offered. Internally, `network1` contains two linked subnetworks: `network2` and `network3`. The external terminations of `network1` are directly mapped to the external terminations of `network2` and `network3`.

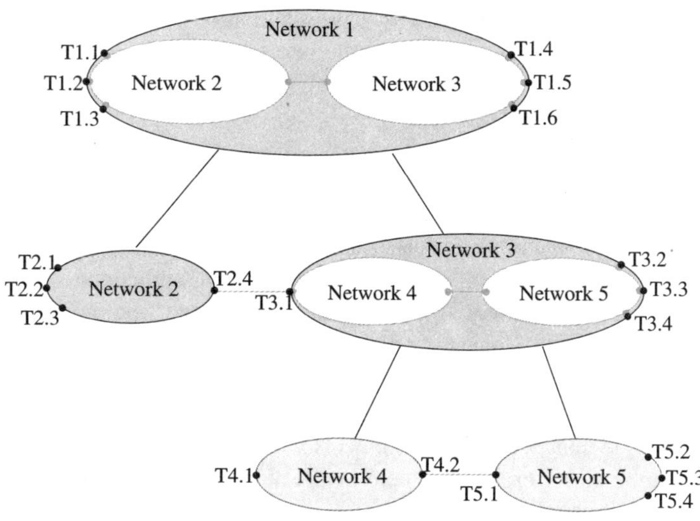

Figure 9.5 Network hierarchy.

Network2 and network3 each publishes four terminations which are made visible to network1. One is used by their super network network1 as the link termination between them.

Internally, network3 has two subnetworks, network4 and network5. All external terminations in network3 are direct mappings of the external terminations of network4 and network5.

The internal views of network2, network4, and network5 are not displayed in Figure 9.5. This can be interpreted as either they do not contain subnetworks or their internal view is hidden from the network modeling exercise.

9.3.1.6 Applications of Topology Management. As indicated before, the value of this topology model is its ability to model and support integrated network management functions. Using this model, any complex network with a specific network technology are modeled as a subnetwork which manifests its services and capabilities through a set of terminations. Mapping these terminations to a super network achieves abstraction of subnetwork services and encapsulation of subnetwork details.

The services provided by the super network is an abstraction of services provided by all of its subnetworks. This abstraction can be used to hide the ways different network technologies manifest their services. In software solution engineering, this abstraction can be used to achieve technology integration. For instance, operations supported by the super network can be implemented in CORBA and operations supported by the subnetworks can be implemented in OSI. The mapping between terminations is used to support CORBA/OSI integration.

The link object between two subnetworks can be used as a cross point from one vendor solution to another, or from one network technology to another. That is, an implementation of a link provides us with a network technology integration point. This allows the end-to-end view of network functions to be constructed which spans multiple vendor solutions and network technologies.

The vertical (termination mapping) and horizontal (link) integration capabilities of this topology model can be developed by using a modular approach to construct network management systems. These capabilities allow each subnetwork management functions to be developed in a truly object-oriented way: autonomous, collaborative, encapsulation, and abstraction. This approach encourages the integration and interworking between different network components, and the solution development is well supported by the component-based software development paradigm.

9.3.2 Generic Connection Model

A generic connection model is a subset of the generic network model. It is a framework for managing network connections between telecommunication systems. It uses the network topology model to maintain the (static and permanent) topology of the network, and it models the dynamic and transient topological concepts. As part of the generic network model, the connection model contains the following basic components:

- Basic connection service objects and their operations. These operations can be used by connection management applications to create, delete, and change network connections.
- A set of semantic rules used to define the behavior of the generic connection model. An example of such rules is the relationship between the generic connection model objects and the network topology model objects.

The connection model contains a number of objects offering connection services. These services are at a lower abstraction level (i.e., network connection level). These services are normally not visible to the end-customers. Instead, they are used to support end-customer service requests relating to setting up end-to-end network connections.

In the generic connection model, the main service objects are:

- a `connection` object—this object represents a connection between a number of terminations within one network. It is the representation of a communication channel provided by a network between a number of terminations. It is responsible for the transparent transfer of information between these terminations. Each network can have many connections.
- a `linkConnection` object—this object is a logical representation of connections created over a link between different subnetworks. Each link can support multiple link connections.

Different from the objects in the topology model, connection and link connection objects represent dynamic concepts. They can be created and deleted dynamically by users according to the application requirements.

Another major difference is that while a connection represents part of the network structure, they always carry service information with them. For example, when a connection is created, the service parameters required by the connection must be clearly specified. Examples of the service parameters include the:

- underlying communication protocol required, such as whether to use ATM or Frame Relay
- bandwidth of a connection
- quality of service parameters required for a connection
- type of a connection, such as whether it is a unidirectional or bidirectional connection; in case of a bidirectional connection, whether it is symmetric or asymmetric

A connection model relies on network topology objects to provide connection management services. The following object relationship between connection objects and topology objects is also depicted in Figure 9.6:

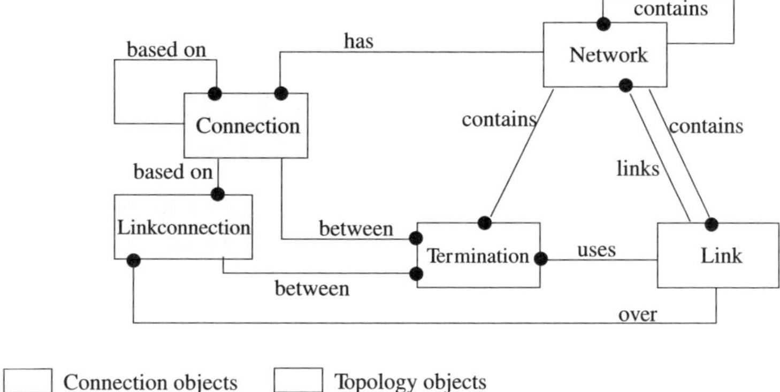

Figure 9.6 Basic connection object model and relationships with topology model.

- A network has a number of connections.
- Each connection is created between a number of terminations.
- A connection in a super network is a composition of a number of subconnections in subnetworks and link connections over the links between subnetworks via their terminations.
- Multiple link connections can be created over a link.

9.3.2.1 Interaction Model. A connection object represents a connection service provided by a network. The generic connection model interacts with the network topology model in the following ways:

- A connection is always totally contained within one network.
- A connection can be created between a number of terminations within one network. Normally a connection contains two terminations, but for a multipoint connection, it may contain more than two terminations.
- A connection in a super network is an abstraction of a number of linked subconnections (in the subnetworks connected by the links). The terminations of the connection in the super network are the abstraction of the terminations of the subconnections in the subnetworks.

Figure 9.7 illustrates connections through terminations within a network.

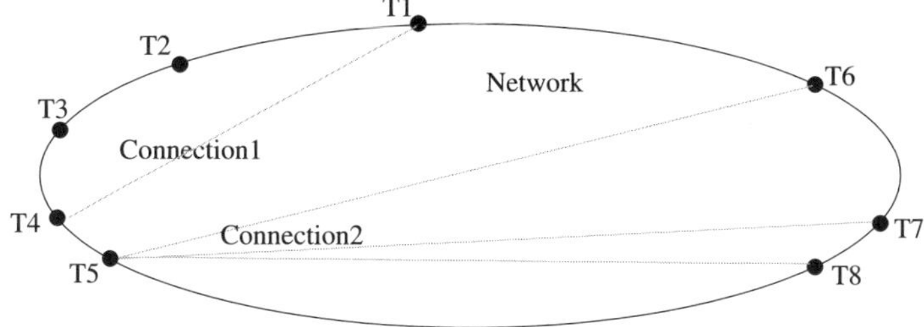

Figure 9.7 Network, terminations, and connections.

Two connections are created in the network. Connection1 is a two-termination connection connecting T1 and T4. Connection2 is a multipoint connection connecting T5, T6, T7, and T8. Other terminations in Figure 9.7 are free terminations.

9.3.2.2 Connection Creation Scenario. The most important operation defined on a connection object is the connection creation operation. Figure 9.8 presents an example network structure. We discuss the connection creation operation based on this example network.

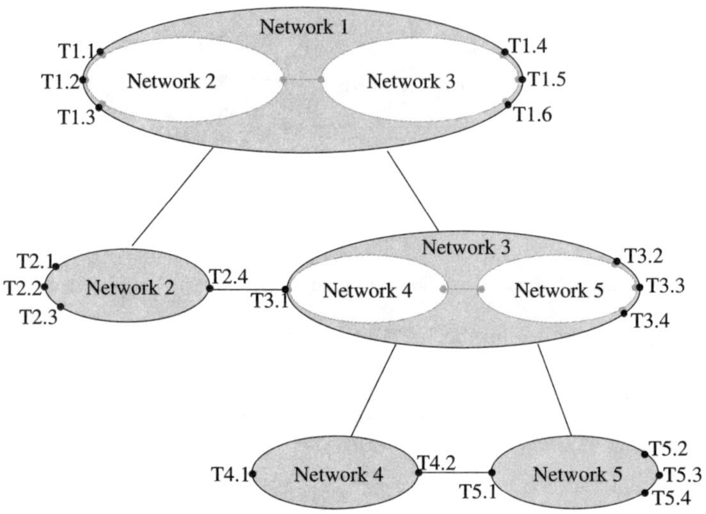

Figure 9.8 Connection creation scenario.

A connection creation operation sets up a connection between the termination T1.1 and T1.5 of the network1. Because network1 is a logical representation of interlinked subnetworks 2 and 3, the required connection is in fact a concatenation of a subconnection in network2, a subconnection in network3, and a link connection between them. The termination T1.1 in network1 is a direct mapping of termination T2.1 in subnetwork network2, the termination T1.5 is a direct mapping of termination T3.3 in network3. The link exists between termination T2.4 of network2 and T3.1 of network3. Two subconnection creation requests are issued by network1 to its subnetworks, which set up:

- A connection between T2.1 and T2.4 of network2
- A connection between T3.1 and T3.3 of network3

Also a link connection is created by network1 between T2.4 of network2 and T3.1 of network3. Figure 9.9 illustrates these created subconnections and link connection.

Since network3 also contains two linked subnetworks, a connection in network3 is also a concatenation of subconnections in its subnetworks and the link connection between these subnetworks. A connection between T3.1 and T3.3 is thus composed of:

- a subconnection between T4.1 and T4.2 of network4
- a subconnection between T5.1 and T5.3 of network5
- a link connection between network4 and network5 via T4.2 and T5.1

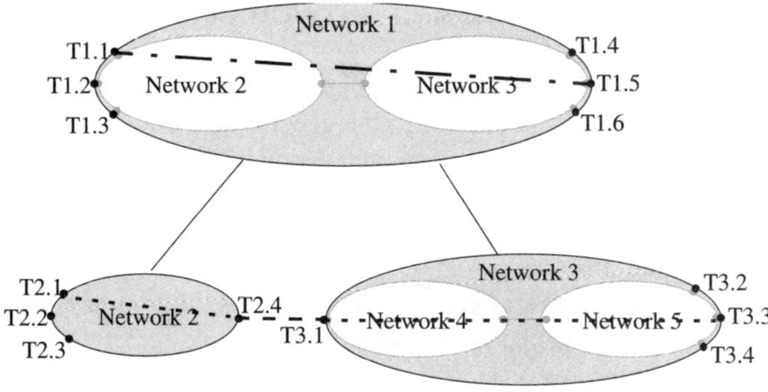

Figure 9.9 Subconnections (1).

Figure 9.10 illustrates these connections.

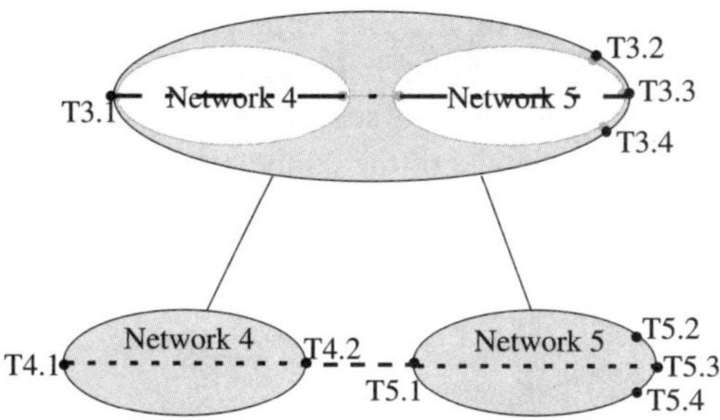

Figure 9.10 Subconnections (2).

Figure 9.11 depicts the final composition of the required connection between `T1.1` and `T1.5` in `network1`.

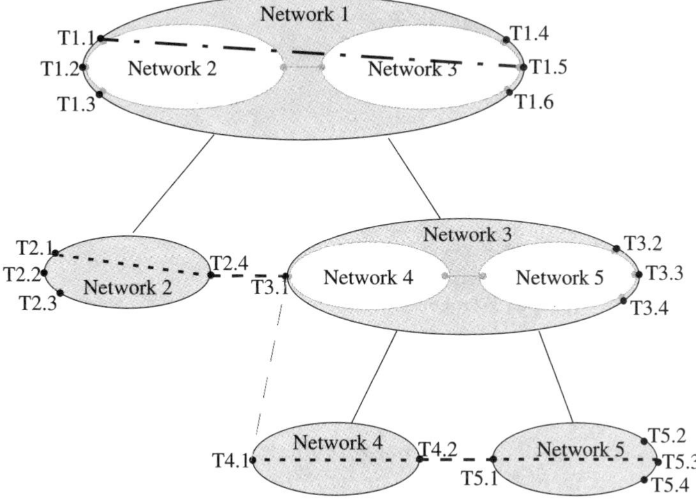

Figure 9.11 Connections created in different networks.

9.4 INTEGRATION OF SERVICE AND NETWORK MODELS

As indicated before, the most important purpose of the generic network model is to support the service management functions. In the generic network model, the concept of connection is not a representation of the physical connection within a specific network technology, it represents an abstract concept of connection in the networks between two termination points. That is, the model supports the end-to-end view of the network topology independent from the individual network technologies.

The generic network model also demonstrates the approach to use the subnetwork concept as an integration vehicle to achieve the interworking between different technologies. In a multiple technology environment, each subnetwork can be designed to represent a portion of the network with one particular technology. An encompassing super network can be seen as representing a technology neutral network topological view. The concept of links provides the glue to connect together two different types of subnetworks.

The generic network model supports the service management functions in two aspects. First, it provides a simplified view of the network functions to support services. This simplified view is crucial to develop end-to-end service management applications in a global, scalable network infrastructure. Revisiting the VOD example in the previous chapter, we can see the VOD service is using object `VODConnection` to support the provisioning of videos in a multitechnology environment. Figure 9.12 depicts the scenario of the end-to-end connection management over a potentially multitechnology network to support VOD service provisioning. In this example, the VOD service session manager has no (nor does it need to have) knowledge of the underlying details of the connection. Thus, it is more independent from the network functions supporting it.

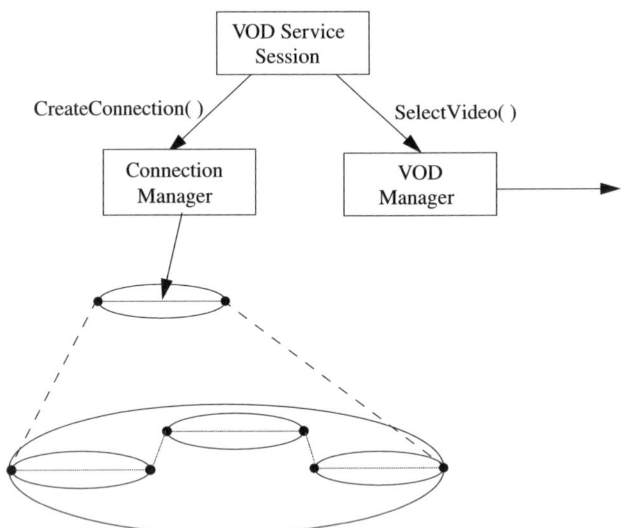

Figure 9.12 End-to-end service provisioning with generic network model.

Using the same principle, the performance model, billing model, etc. can be developed as part of the generic network model. The VOD service management system could use these integration models to support end-to-end customer quality of service management.

The second area generic network model can support service management is by providing an environment for multiple network technologies to interwork. As the generic network model contains only the most general concepts and interactions among the many network concepts it is likely that this model represents a minimum yet sufficient manifestation of the network function to support services. This increases its chance to be an integration platform for different technologies.

Also the integration approaches discussed in the previous chapter can be applied to the integration of these models in a system where different management technologies are used. For example, the IDL-GDMO mapping approach can be used to implement a link. This allows two interlink subnetworks to be managed by different software technologies. The abstract mapping approach can be used to support mappings between external and internal views where the semantic abstraction is required. The application integration approach can be used to support super network composition from subnetworks. This allows the application level end-to-end view.

The simplicity of the model makes it possible to view any technology-specific network models (ATM, SDH, ADSL, WDM, etc.) as the multiple specializations of the model. This generalized model always provides a better integration framework and a more stable interface support for the applications developed on top of them.

In the next section, we discuss how this generic network model can be used to integrate the specific network models and network management systems based on them.

9.5 INTEGRATION WITH NETWORK MANAGEMENT SYSTEMS

The purpose of defining a generic network model is to provide a simple and abstract view of the network and network services to support the service layer. The generic network model is not a new network protocol. It is an integration device with the network management functions existing in the specific network models and implemented network management systems.

One key concept of the generic network model is the mechanism for abstraction. A super network is not a simple collection of a set of subnetworks. It provides a layer of abstraction of the subnetworks functions. The abstraction includes the network structure, user access points (terminations), and the services and their characteristics. This abstraction allows the generic network model to hide the complex details existing in the network management layer as these details are not relevant to the service management layer. It also encapsulates the heterogeneity of different underlying networks and their management systems.

There are two important aspects when discussing the relationship between the generic network model and a specific network model: the *abstraction* and the *specialization*.

The abstraction is an activity performed during the network modeling stage. It is to provide an abstract view for the specific network to be managed. A logical network is created using generic network model, which retains all the knowledge obtained during the object modeling stage. Examples of such knowledge include:

- The number of subnetworks contained in this logical network and the way they are connected
- The external view of each subnetwork, in terms of external terminations and services supported through these terminations
- The types of connections that can be set up in each subnetwork and the characteristics, more importantly, the abstract view of these characteristics in this super network

The specialization is an activity performed as part of network management implementation. It specializes each management request (in the logical network modeled by generic network model) to a set of specific management requests with a set of specific parameters in its subnetworks. The knowledge obtained by the generic network modeling process during the abstraction stage is used as guideline for the specialization. This abstraction and specialization relationship is illustrated in Figure 9.13.

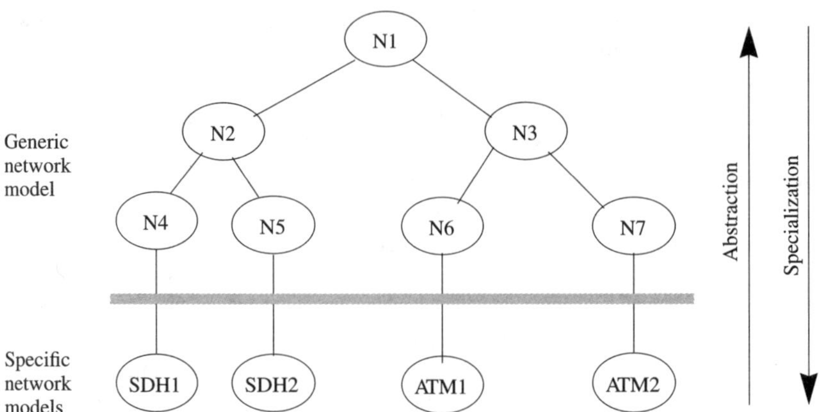

Figure 9.13 Specialization of the generic network model.

A logical representation of each specific network can be created in generic network model, indicated in Figure 9.13 as N4, N5, N6, and N7. A logical network understands the external view of the specific network it is representing. Especially, it has the knowledge of the external terminations, that is, the service access points of the specific network. If a logical network represents only one specific network, then the external view of the logical network reflects the abstraction of the external view of the single network. If a logical network contains more than one specific network as its subnetworks, then the external view of logical network represents the linked specific subnetworks as a single network to the outside world. The internal view of the logical network includes the external view of each of the specific subnetworks and the topology links between them.

The networks in generic network model can be further grouped to represent higher level abstractions. In the example of Figure 9.13, N2 is a logical network containing N4 and N5 and it represents an SDH management domain. N3 is a logical network containing N6 and N7 and it represents an ATM management domain. Again N2 and N3 can be grouped together to form N1 which represents the management capability in a cross-domain network.

9.6 STANDARDS CONFORMANCE

Strongly related to the generic object modeling is the issue of standards conformance. In fact, ITU-T network model standards defined in GDMO were developed within the context of the OSI framework being used for achieving network management interoperability.

Conformance to standards in TMN management world appears always to be a requirement. However, it is useful to explore this conformance issue further.

Conformance deals with interactions between different components built for different purposes or by different people. We differentiate two types of conformance for different purposes: for *protocol interoperability* and for *solution migration.*

Conformance for protocol interoperability is easier to understand. Two components of the system exchange bits and bytes in a form that both components can recognize. The current network management technology and information models are primarily aimed to achieving this.

We are more interested in solution life-cycle management in a multiple technology environment. We are more concerned with how solutions evolve at minimum cost as the underlying technologies evolve. Although the theory is that standards conformance helps the migration of solutions, our experience is that to harmonize the interactions of two components is more important and useful than to conform to a standard. To achieve harmonization, we first define the terms *maximum specification* and *minimum specification.*

When we build a component, we should always consider that we are building a specialization of a much larger solution. One way to achieve this is to fully specify its interface and behavior. When the components we built are used by other people to create a bigger solution, they understand exactly what we were trying to do—that is, we always anticipate that someone will generalize our component for other purposes. This is always a good practice to avoid building legacy systems.

When we integrate with another component, we should always consider a minimum specification for this integration—that is, we treat the component we are going to integrate as a generic component. This allows us to access the generalized functionality of the component, yet reduces our dependence on it. This is always a good practice to support the application migration and evolution.

Thus, for every software component, two specifications exist. A maximum specification defines what you are building, and a minimum specification defines what you are building for. The maximum specification defines the operational behavior of a component, and the minimum specification defines the interaction behavior of the component.

An example is the interaction of buying a car: a car producer will always provide a full specification of the car (all the way down to the size of the bolts), while a consumer always provides a minimum specification when buying a car (such as color and number of seats).

In network management, network models, when fully specialized to a particular technology (such as SDH/SONET), provide a maximum specification which is used for protocol interoperability and easier technology migration. The generic network model, on the other hand, is a minimum specification of very high level functions for service management applications to interact with. These interactions will be minimally impacted by technology migration (for example, from SDH/SONET to WDM).

This is particularly important when we integrate with a component designed using an older paradigm. An example in the TMN solutions area is a CORBA-based service management application integrating with an OSI network management system. The minimum specification ensures that we don't try to implement the old paradigm (OSI) using the new language (CORBA). Instead, we generalize the OSI component using the new paradigm.

The approaches described in this chapter reinforce this principle. The specialization of the model supports detailed maximum specification for the purpose of software migration. We know exactly the details of the system and its components. The generalization of the models is minimum specifications for the purpose of integration. The service model only manifests the very basic business rules we interact with the customer; and the generic network model only specifies the most generic concepts the network supports.

As it is neither necessary nor possible to standardize the business operation environment for all telcos in the world, it is not the aim that the service and network model be standardized and adopted by everyone. It is important, however, that every telco models their individual business environment from the business viewpoint, unfettered by the underlying network technology and software solutions for managing networks. This provides hope that business solutions based on this model will have a longer life cycle than the management technologies, and that telcos will not have to reengineer their business at the pace of technology migration.

Standard conformance is a means, not a goal. The goal is to have a solution which supports the business and offers value to the customers. This solution needs to survive the rapid change of the business and technology environment. In doing so, we need to constantly and dynamically interact with the outside world and within our own environment. Standard conformance may help us, in some occasions, to achieve the smooth interactions. This is all the value standards can offer. It is ultimately our responsibility to ascertain their value and decide our strategy. More importantly, the standards process is also a constant evolving process; as such, it is also one of the external risks our solution has to manage.

9.7 SUMMARY

The object modeling exercise described in this chapter focuses on a process which centers on application interaction and integration. This process illustrates that while operational specification for functional components is important, the specification of the interactions between components are equally, if not more, important. Referring to the integration purposes and integration approaches discussed in the previous two chapters, the models discussed in this chapter can be characterized by the Figure 9.14 to indicate their roles in integrated architecture.

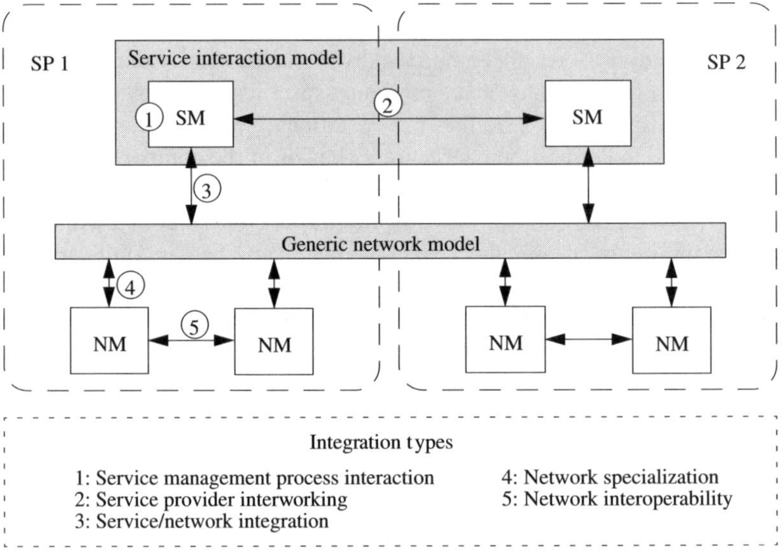

Figure 9.14 Generic object models and their roles in integrated architecture.

It is useful to stress that the object models presented here do not necessarily correspond to layered software architecture with extra software components. The service interaction model and the generic network model only represent different views derived from the specifications representing service management functions and the network management functions. They are minimum specifications for interaction purposes. These specifications can be supported by the same software architecture supporting operational specifications.

10

Integrated Management Architecture

This chapter presents an integrated management architecture for building distributed network and service management solutions. This architecture can be used as an evolving platform to define, deploy and manage customer and business solutions.

The focus of this architecture is on how to support telco business solutions and to deliver the services to the customers. It focuses on the integrated business processes and information flow-through. This architecture addresses the business requirements and objectives for the integrated management. It supports the definitions of management functions and business processes and the interactions between them. The architecture presented in this chapter is not focused on technology choices. In the next chapter, we discuss the issues of mapping this architecture to a software architecture to implement the management functions defined here.

10.1 INTEGRATED ARCHITECTURE

An integrated management architecture is depicted in Figure 10.1. It contains the following three environments:

- *A customer access management environment*—It offers the functionality for customers to access telecommunication services. This environment contains two major components:
  —a set of interfaces to enable end-customer access to telecommunication services using different means
  —a customer care management component which manages all the customer access and interactions.
- *A service management environment*—It offers a set of service management functions and a set of business processes. It is a middle layer between customer management environment and network management environment. On the one hand, it presents the network capabilities in terms of services which can be understood by customers. On the other hand, it relates customers' service requests to a set of operations which can be understood by network functions. This environment has four major components:

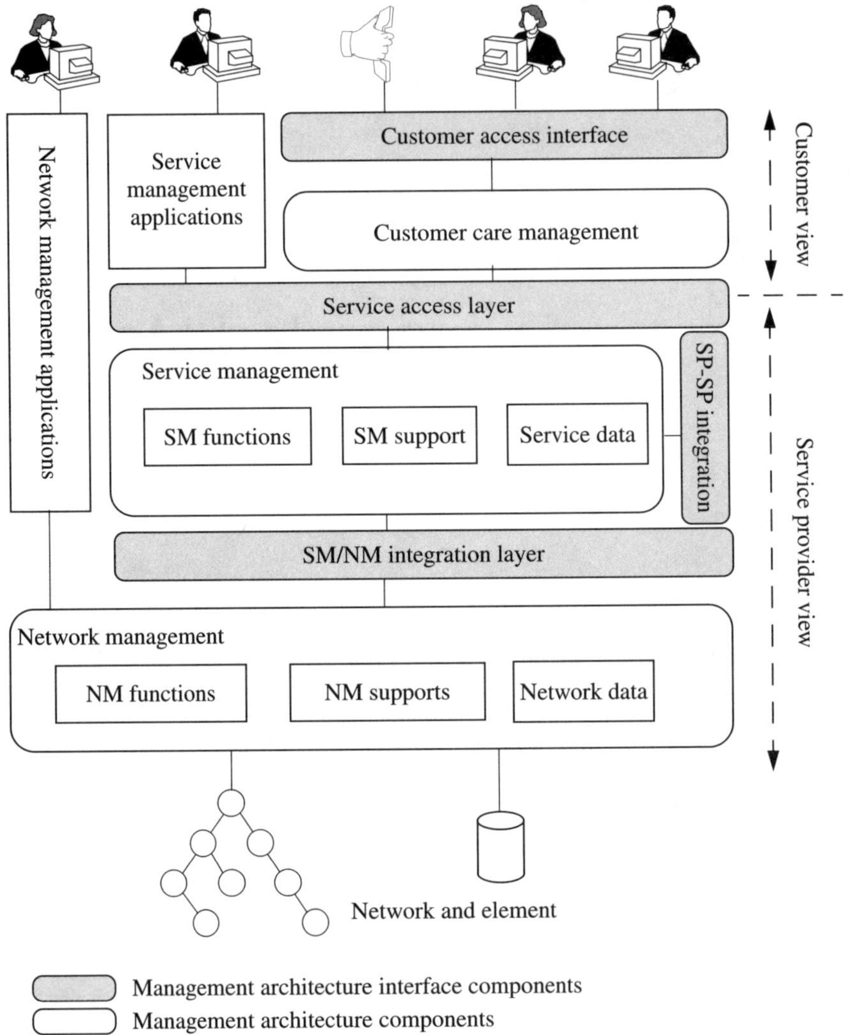

Figure 10.1 Integrated management architecture.

—a layer of service access interface to support customer management system and service management applications
—a set of service management functions and supporting functions
—an integration component to manage the interworking to other service providers
—a service and network integration layer to support the integration with the network management functions

- *A network management environment*—It is an environment for managing networks and network elements required to support the services to end customers.

The layers in the architecture reflect different views in the different levels of object abstraction. The managed environment (including elements, networks, services, and customers) represents continuous steps of abstraction upwards. In this abstraction process, the managed environment manifests different views. The service access layer and customer care layer reflect the customer view of the service provider environment. The rest of the architecture reflects the service provider's view. It is further partitioned into two components, one represents a product and service view, the other represents a network resource view.

This architecture is extensible. In terms of functionality, more functional blocks can be added to different environments to increase the management capability and scalability. In terms of application development, different telecommunication business processes can be implemented using the customer access environment, service management environment, and directly using network management environment. In terms of managed network, multitechnology transmission networks can be added with little impact to the business processes already implemented.

This architecture can also be scoped down by including a subset of the functionality in each management environment to meet the minimum telco management requirement.

Different technologies can be used to implement different components in these management environments. Distributed OO technologies, such as CORBA, can be used as the platform to support the integration between environments, and between different components. The rest of this chapter and the next chapter discuss these issues in more detail.

10.2 CUSTOMER MANAGEMENT ENVIRONMENT

The customer management environment provides the following functions:

- Customer access to the OSS using various GUI technologies
- Customer interaction through management processes
- Customer management architecture

These functions facilitate customer interactions with service providers for the purpose of:

- accessing customer account information, such as service portfolio, bill presentment, enquiries and payment.
- supporting sales process, such as searching for a products and service catalog, customer relationship management, sales and promotion and on-line service ordering and provisioning.
- supporting customer satisfaction management, such as managing trouble reporting, QoS reporting and monitoring, problem prediction and resolution.

The customer management environment supports various means of interaction between customers and service providers' OSS environment. It presents TMN service and network management functions to end customers with a viewpoint that makes sense to them. This viewpoint may not correspond to the traditional service providers' viewpoint as defined by FCAPS functions.

As such, the customer management environment usually does not require the change to the existing OSS systems; instead, it is an architecture that integrates with the existing OSS systems. The architecture supports a set of customer access interface technologies, a set of customer management business processes, and a set of business objects. These business processes and objects represent the underlying OSS functions and status.

The customer access architecture allows customers to use various access interfaces to invoke customer management business processes. These processes interact with the core OSS and service management systems through a set of business objects.

The architecture supports three layers of components and processes. First, it supports various interface technologies that end customers use to access the service management environment. These technologies may include:

- Web browser
- Interactive voice response (IVR)

- E-mail
- Java client
- Information kiosks

Second, this architecture supports various business processes enabling customer access. These business processes may include the following on-line access activities:

- Search product and service offers
- Order services (teleconference, video, leased line, etc.)
- Review account portfolio
- Review billing information and initiate payment
- Review SLA information
- Access to virtual private networks (VPN)

Third, the architecture supports various business objects which represent the element of service offers and management capabilities. These business objects include:

- Service level agreement
- Service order
- Teleconference service
- VPN
- Video connections
- Billing records

The customer access architecture supports the mappings of these business processes and business objects to the existing OSS core systems (billing systems and databases, customer database, service catalog database, provisioning systems, and broadband management functions, etc.).

10.2.1 Customer Access Interface

A customer access interface is the customer's interface to an application. A customer's session with an application may involve the use of multiple customer access interfaces, for example they might place a request via a Web page and receive confirmation via a fax.

The customer management environment provides an architectural framework for the development and execution of customer access interfaces. This framework ensures that customer access interfaces are capable of readily adopting new customer access paradigms and technologies.

The customer access architectural framework is capable of supporting new and existing user access technologies. Existing user access technologies may include: screen, Web browser, phone, e-mail, conventional mail, and fax. It is intended that the framework addresses all forms of customer access, not just the electronic ones (as electronic customer access does). In addition, the customer access architecture should allow new customer access technologies to be readily introduced to an application alongside existing ones. This introduction of new customer access technology may require new customer access business processes and objects to be developed, but it should not affect the existing business processes and business objects.

At the heart of the customer access interface is an interaction model of the high-level business processes that the customer can have with the system. Ideally, this model remains constant as the paradigms and technology around it evolves to meet the application's ever-changing requirements. To accommodate this ideal, the customer access architecture needs to isolate the part of the application that is affected by changes in the paradigms and technology from the part of the application that endures through such changes. The part that is affected by changes in the paradigms and technology can be broken up into smaller parts based on their purpose and the types of technology that they use. These parts can also be isolated to minimize the amount of change that a new paradigm or technology may cause.

The overall customer access system architecture focuses on the integration of multiple user interface access technologies and existing service layer functionality. This architecture is shown in Figure 10.2.

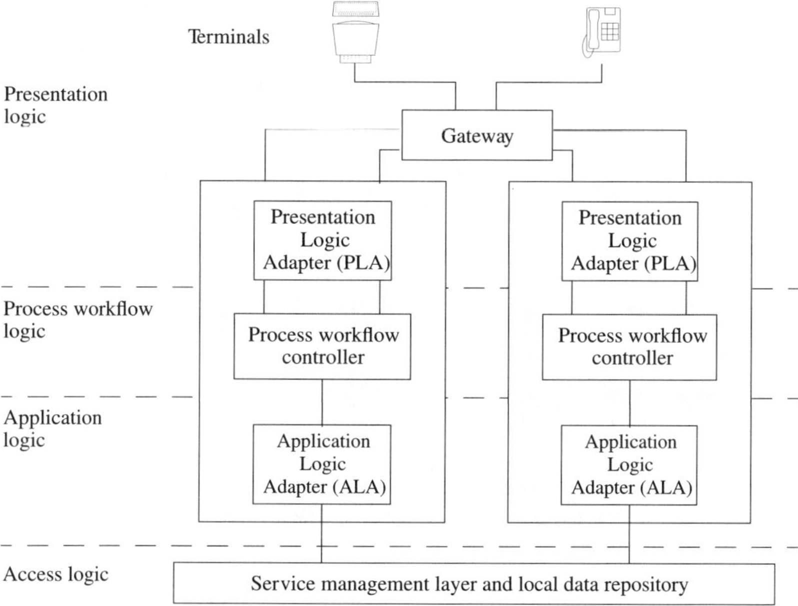

Figure 10.2 Customer access systems architecture.

Four major components are defined in this architecture:

- *Presentation logic*—This provides the framework to handle and integrate multiple GUI technologies with the customer access framework.
- *Application logic*—This provides the support of business objects which represent the semantics of the services that customers are requiring.
- *Access logic*—This provides the access to the service management layer and its local customer access record required to support the customer access semantics.
- *Process workflow logic*—Process workflow logic component provides the linkage between the application logic to the presentation logic and the access logic. It forms the core of the customer access systems.

The Model-View-Controller (MVC) customer interface paradigm is used in this architecture. A customer access interface is partitioned into three parts:

- the model, which provides the core semantics of the user interface's functionality
- the view, which is concerned about how the information is presented and accepted
- the controller, which supports the model by interacting with the underlying system

The architecture has three tiers and each tier addresses a particular part of the MVC access interface paradigm.

The presentation logic tier addresses the view part. It facilitates the existence of multiple user access technology specific views. Introduction of a new customer access technology should only require that a new view be added.

The process workflow logic tier addresses the model part. It provides a layer in which the high-level business processes representing application semantics can be realized. It is concerned with things such as where the customer's session is positioned in the application's menu hierarchy, controlling the customer's access to the application's operations, and auditing the customer's use of the application's operations.

The business processes and application logic tier provide the controller part and the underlying facilities that the controller works with to provide the application's functionality.

Two separate applications are shown in Figure 10.2. Each application is represented by a grey box and comprises of a set of presentation logic adapters (one for each of the user access technologies required by the application), a workflow controller, and an application logic adapter.

The following architectural components are identified in Figure 10-2:

- Terminal
- Gateway
- Presentation logic adapters
- Process workflow controller
- Application logical adapter
- Access logic

TERMINAL. A terminal is an I/O device that customers use to interact with the application. It can be a dumb terminal, a phone, a personal computer running a Web browser, a fax machine, an e-mail box, etc. The protocol used by the terminal to communicate with the application is known as the user access protocol.

GATEWAY. The gateway component is the customer's first point of contact with the system that supports the application. Its main purpose is to filter out communications from invalid system users and to ensure that the channel between the terminal and application satisfies the application's security requirements. Its secondary purpose is to act as a switch that connects incoming communications from a terminal to the requested applications. It binds an incoming request from a terminal to a relevant presentation logic adapter and outgoing communications from a presentation logic adapter to the associated terminal.

Gateways are stateless, and customers are able to access an application via any of the system's gateways that support the required user access protocol. Gateways may support multiple user access protocols.

The stateless gateways in the software architecture can increase the scalability and distribution of the system. The customer may need to use a directory (or a CORBA trader) to determine which gateway supports the required application and the user access protocol.

PRESENTATION LOGIC ADAPTER. This adapter component translates the customer requests and the responses sent by the terminal into a form that the workflow controller can understand. It also handles the information presentation layout that is sent to the customer's terminal. For example, the HTML pages sent by an application to a Web browser terminal will be prepared by the presentation logic adapter. Another example is that the voice prompts being sent to a telephone terminal are also prepared by the presentation logic adapter. There are cases where the terminal is solely responsible for the presentation layout, for example, a java application that uses the IIOP access protocol. In these cases, the presentation logic adapter has little or no influence on the terminal's presentation layout.

PROCESS WORKFLOW CONTROLLER. The process workflow controller controls the execution of the business process instances. It decides which actions a customer can execute and the order in which they are made. Access control restraints are used to control who can initiate an action, and a Finite State Automata (FSA) is used to control their order. The FSA requires the workflow controller to keep track of the state of each of its customers' sessions.

The workflow controller receives requests from the presentation logic adapters and returns to them the corresponding replies. It also forwards valid requests onto the application logic adapters and receives from them the corresponding replies. Since a presentation logic adapter and application logic adapter are specific to a particular workflow controller, the protocol that they use to communicate with the workflow controller can be selected by the designer of the workflow controller. In most implementations, it is likely that they will share the same process and thus communicate via local function calls.

The process workflow controller is capable of auditing the high-level actions made by customers.

APPLICATION LOGICAL ADAPTER. This component performs adaptations between the customer access interface and customer care management processes, which take care of business and application logic. The functionality of this area will be discussed more as a separate system component: a customer care management process.

ACCESS LOGIC. This component supports the access to the OSS and other services and network management functions.

10.2.2 Customer Care Management

Customer care management processes provide support for customer access to the OSS systems. Its main functionality is to support the customer access interfaces discussed in the previous section. The customer care management process is responsible for:

- Managing quality for customer ordered services
- Interacting with customers via customer access interface
- Interacting with service management processes
- Managing customer access security

The customer care system manifests the status of the service management processes to the end-customers through the customer access interfaces. It handles customer enquirics for new services or presents the status of their existing services. It ensures that each customer request relating to the services is properly tracked by relevant service management processes. The customer care system manages the whole life cycle of customer requests until responses are provided to the customer.

All interactions with customers are made through the customer access interfaces using customer preferred interface technologies such as Web, e-mail, telephone, etc. with the support of the presentation logic adapter. The customer care manager is responsible for initiating the interactions, such as:

- Invoice is ready to be send to customer
- Performance reports or SLA reports are available
- A service alarm occurred which may affect some customers' services

Customer care manager may not understand the internal processes of service management but it has the knowledge of the interfaces supported by the service interaction model. The customer care manager issues service management requests on behalf of customers. These requests invoke service management process functions defined in the service interaction model.

Figure 10.3 illustrates a scenario of customer interaction for service ordering and billing enquiry. The figure shows the flow of processes and information between customers to customer care systems and to the internal service management processes.

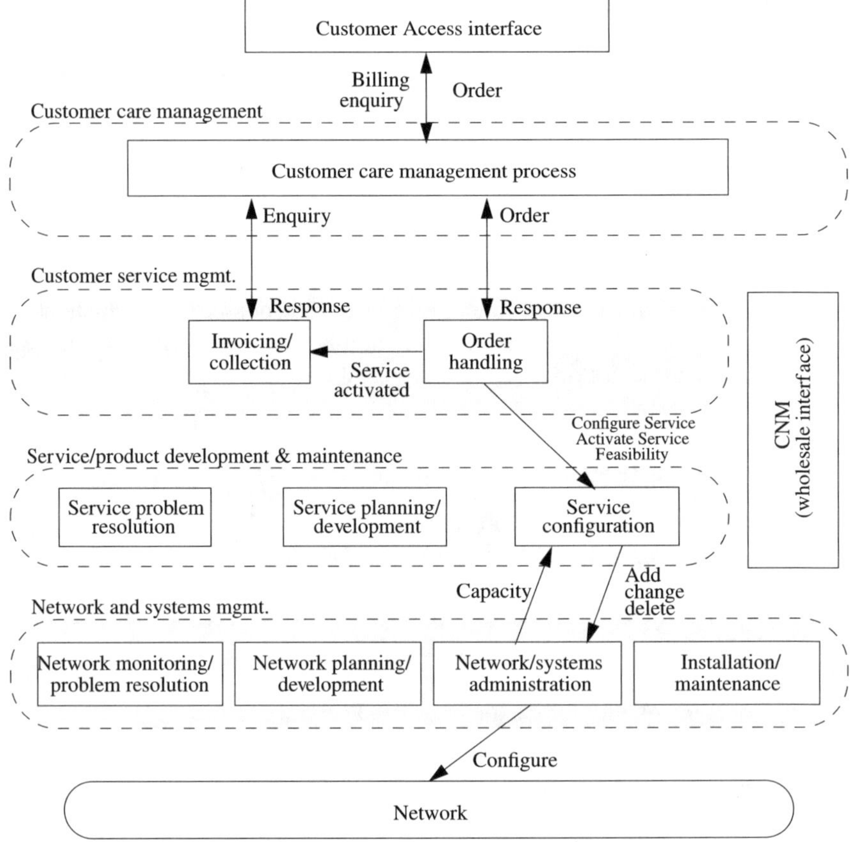

Figure 10.3 Customer care view of service ordering and billing enquiry.

The customer access interface is used by the customer to formulate both billing inquiries and service orders requests. These requests are then sent to the customer care management process. The customer access interface and customer care management processes are responsible for the first level of security and identification of customers. Validation and identification of customers will be important in ensuring privacy of transactions. Having identified the customer and ensured that the customer has sufficient privileges to make this request, the request is forwarded onto either the invoicing/collection or order handling processes.

The invoicing/collection process then finds the appropriate records from the billing database and formulates a response for sending back to the customer.

In the case of ordering request, the service ordering process is activated to supply this order request. The service ordering process eventually flows to all levels of the OSS and the configuration system to fulfill the order in the network.

10.3 SERVICE MANAGEMENT ENVIRONMENT

At the core of the integrated architecture is the service management environment. This environment is composed of two parts: the service management processes, and the process interactions and integrations. The service management environment defines a number of service management processes and these processes support essential service management activities. In this section, we discuss the following processes:

- Service order and management process
- Performance management process
- Problem management process
- Fault management process
- Billing process
- SLA management process

The service management environment also defines a number of interface components supporting interactions between service management processes and integrations with system components outside the service management environment. These interface components include:

- A service access layer supporting service access interfaces which can be used by customer management layer for managing customer interactions
- A service and network integration layer supporting the integration between objects models representing service management and network management functions. This integration layer associates customer service management requests to a set of network management requests.
- An SP-to-SP interworking component supporting communications with other service providers for service management functions.

Figure 10.4 depicts the service management environment architecture and the components.

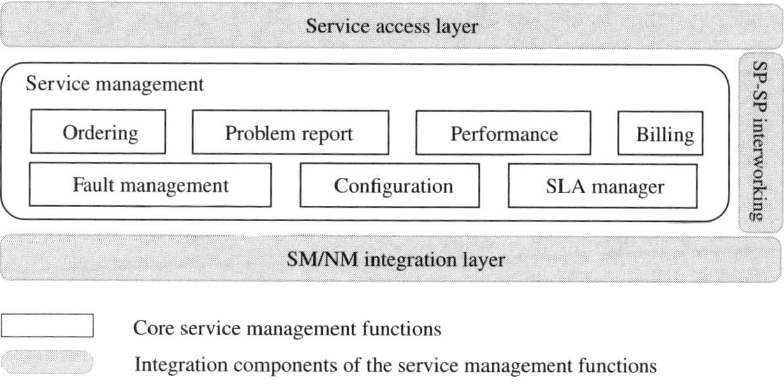

Figure 10.4 Components in service management environment.

10.3.1 Service Management Processes

Service management processes support the core functionality required to provide telecommunication services to customers. These service management processes are defined as functional building blocks in the overall management architecture.

Figure 10.5 illustrates a service management environment consisting of these management functional building blocks and the interactions between these building blocks.

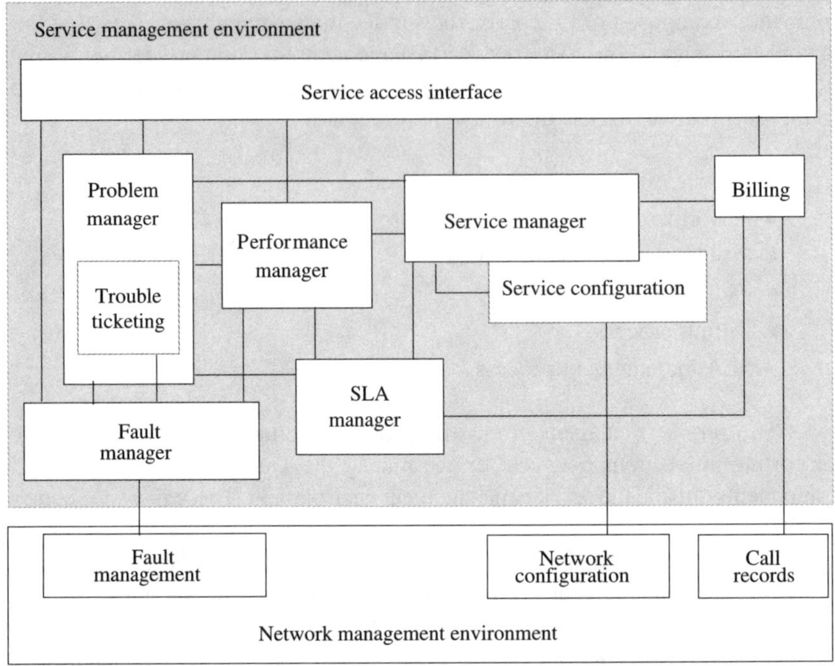

Figure 10.5 Functional blocks in service management environment.

10.3.1.1 Business Process Workflow. The service management processes discussed in this section are represented as process workflows. Each workflow is defined as a collection of tasks organized to accomplish a process in a specific way determined by the business rules.

A task can be performed by one or more software systems, one or more humans, or a combination of both. Human tasks include those with or without direct interactions with computer systems. Examples of tasks include updating a document or a database, generating a report, contacting a customer, and laying a cable (in service provisioning applications).

In addition to a collection of tasks, a workflow defines the business process rules. These rules define the interactions between tasks and form the semantics of the business processes. These rules may include:

- Task invocation order, constraints and conditions
- Task synchronization rules
- The temporal logic between tasks
- Concurrency rules
- Transactional behavior of tasks
- Information flow rules

These rules form the integral part of the business process and information modeling. They are specified at the information modeling stage.

In the service management processes described here, a "task" is a functional component of the service interaction model. A task does not have to be a newly implemented function, it can be provided by existing OSS functions such as retrieving a call record from the billing data collection.

In this section, we use a very loose notation to construct the workflow diagrams representing the service management processes. These workflow diagrams define the instance level interactions between different processes in the service interaction model defined in the previous chapter. Figure 10.6 lists the notations used in this context.

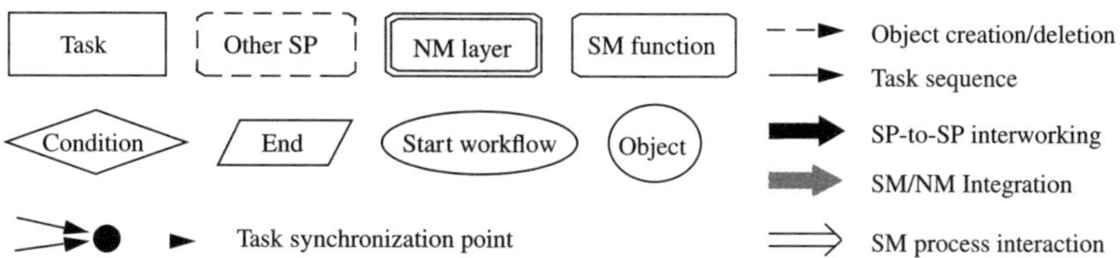

Figure 10.6 Workflow notations representing interaction model for service management processes.

10.3.1.2 Service Manager. The *service manager* supports the generation and provision of a customer service order. It interacts with the service configuring process to require the appropriate network connectivity to supply the service specified in the customer order. The service manager workflow performs the following tasks:

- Determines the feasibility of an order (i.e. whether there are sufficient network resources)
- Configures services to fulfill the order
- Activates the services
- Updates and deletes the services if required (after the services are activated).

The service manager performs these actions by invoking the internal *service configuration* process. Service configuration process is capable of configuring the network and determining network capacity and acts as an integration point between service processes and network processes.

Normally, a service life cycle contains following phases:

- *Sales phase*—customers access the service and product catalog
- *Pre-order phase*—customers interact with the service manager to obtain service offers and options
- *Order phase*—customers interact with the service manager to place a firm order for a service and agree to an SLA
- *Service delivery phase*—service manager delivers the services to customers and manages the SLAs.

SALES PROCESS. During the sales process, customers gain access to the SP's service and product catalog. This customer access can be supported by various means such as Web and telephony. Typically call center and search engine technology can be employed to support the product catalog function.

This process proceeds to the pre-order phase in a retailer customer access scenario. It is part of the customer care function and is normally delegated to the customer care management system.

PRE-ORDER. During the pre-order phase, customers require information about services with the intention to order them. The pre-order process is responsible for preparing and presenting offers to the customers. The pre-order process may respond with multiple options in an offer—each option can have different service features and quality of service parameters.

Figure 10.7 illustrates a workflow of a pre-order process consisting of multiple tasks. The workflow does not intend to define the operation of the process. It focuses instead on the interactions between this process and other service management processes, the interworking with peer SPs, and integration with network management processes.

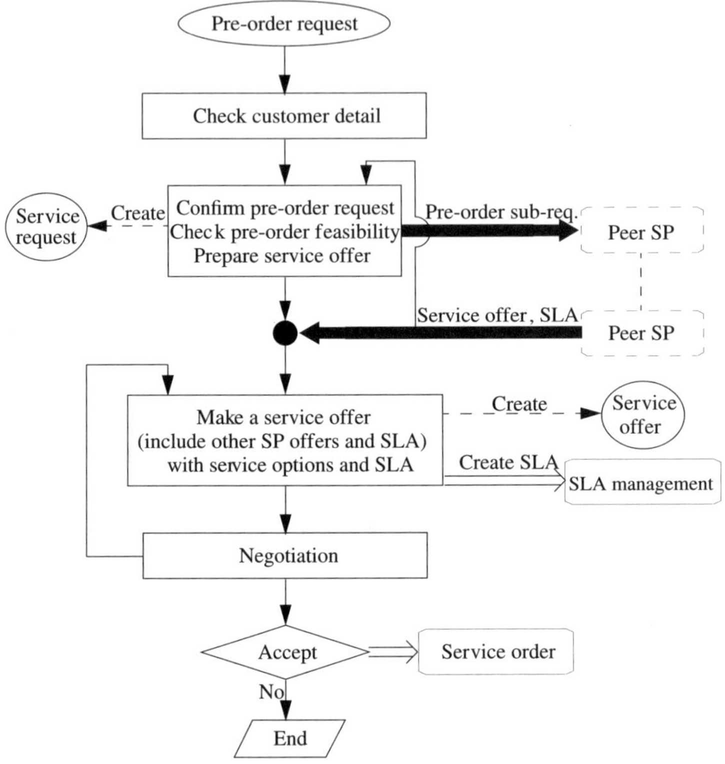

Figure 10.7 Pre-order process workflow.

One of the key activities in the workflow is to prepare a service offer. It could be the case that the service required by a customer cannot be solely provided by this service provider. Other service providers have to be involved to provide the components. In this case, this service provider needs to enter into the pre-ordering phase with other service providers for the service components. The service offers, options, and SLAs from these service providers will be incorporated in a complete service offer and SLA for the initiating customers.

Another activity in the workflow is the negotiation subprocess between the customers and the service provider. This subprocess is used by the customers to negotiate the service features and SLA details. These may include the network bandwidth, service downtime, error rate, cost, etc. The negotiation activity may iterate a number of times before an offer and SLA is acceptable to the customers.

The workflow manages the object life cycle for the service interaction model discussed in the previous chapter. The life-cycle management includes the dynamic creation of service interaction objects as illustrated in Figure 10.7. The workflow also controls the interaction with other service processes supporting the service interaction model as well as interworking with peer service providers.

The pre-order process terminates when the customer is satisfied with the service offer and is prepared to enter into the ordering process. Alternatively, the process may terminate without a satisfactory offer, or the customer may decide to terminate the process without entering to the ordering phase.

Order Process. The service ordering process enables a customer to place a firm service order. This process can be proceeded by a pre-order process. Alternatively, customers can enter into this process directly (or after a catalog search). Figure 10.8 depicts the workflow of the service order process.

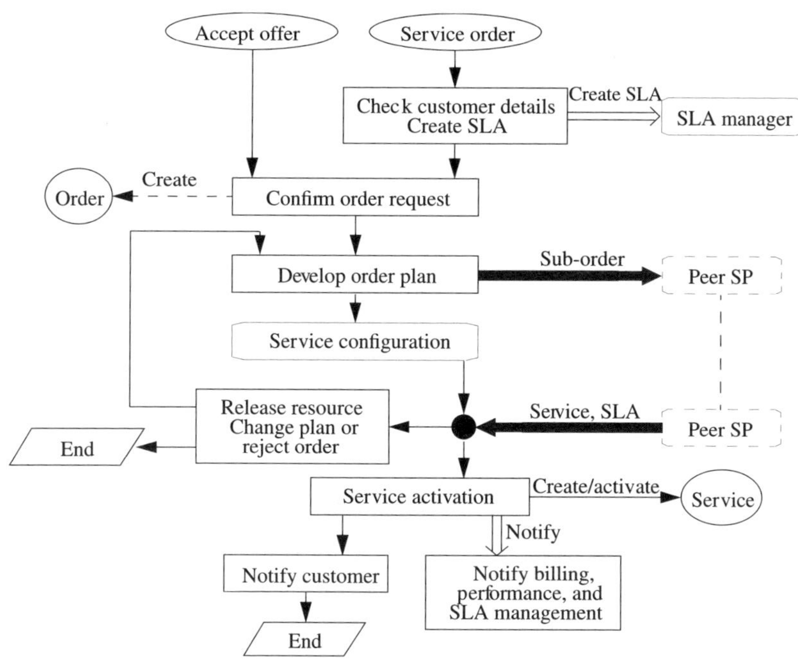

Figure 10.8 Service order process workflow.

The service order workflow controls a number of tasks. One task is to develop an order plan. This task may involve ordering service components from other service providers. Similar to the pre-order process, the service orders and SLAs from these service providers will be consolidated to form a complete service order and an SLA for the initiating customer.

The order process may invoke internal service configuration process for the provisioning of the service. The service configuration process determines the feasibility of a service order. It then configures and activates the service.

The service ordering process also manages the service interaction model and controls various interactions and integration points defined in the service interaction model. These include the interactions with the SLA manager and the peer service ordering processes. The process also interacts with the customer care system.

The service ordering process terminates when the service is activated and delivered to the customer. For a scheduled service, the service is activated according to customer specified time. This signifies the end of the service fulfillment phase. The life cycle defined in the service interaction model moves to the quality of service management.

Configuration Process. Service configuration process is an internal process to support the service ordering and provisioning. It is a key process to support the integration between service management processes and network management processes. This process has the knowledge of both the service interaction model and the generic network model in order to support the process and information flow-through between the two models. Figure 10.9 depicts the workflow of configuration process.

As illustrated in Figure 10.9, the process is the main integration point between service processes and network processes. It achieves this by maintaining the object relationships between two object models. In the technology view of the integrated architecture discussed in the next chapter, we will further demonstrate that this is also a key process to achieve technology integration between different management technologies.

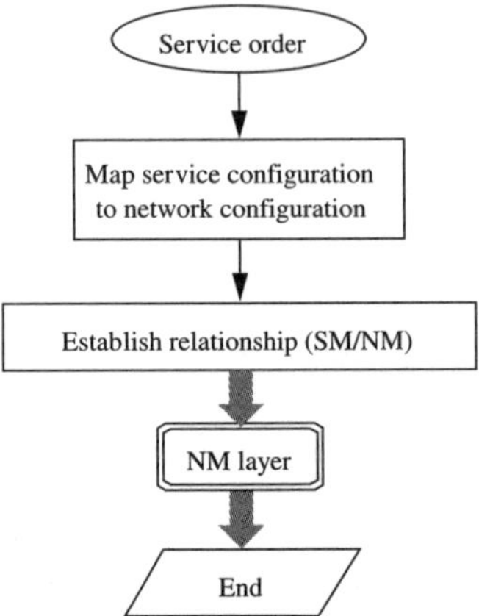

Figure 10.9 Service configuration process workflow.

SERVICE CONTROL PROCESS. After a service is delivered and activated, the customers may wish to delete the service. The delete process releases all resources used for that service and interacts with billing process, SLA manager, and performance manager for the termination of the service. Figure 10.10 illustrates the workflow and activities for a delete service process.

The process interacts with other service management processes. In particular, it integrates with the network management processes by invoking service configuration process to release the network resources. It also interacts with other service providers if some of the service components were provided by those service providers.

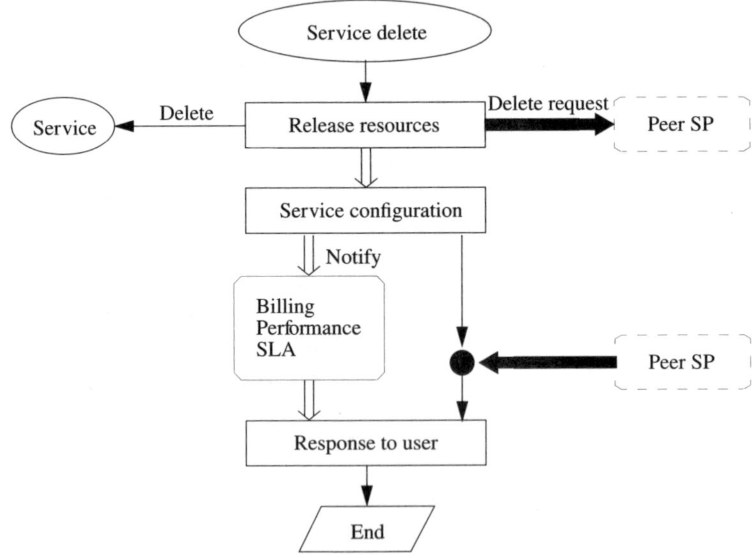

Figure 10.10 Service delete process workflow.

10.3.1.3 Performance Manager. The *performance manager* produces scheduled reports on performance statistics. The reports are generated based on the analysis of performance related data. These data are gathered from multiple sources including:

- Service activation and termination information from service manager
- Potential capacity problems based on network usage statistics from network monitors
- Performance data from the network layer
- Fault/alarm information from fault manager
- Trouble report information from the problem report and handling manager

The performance manager examines these information and reports any developing capacity problems and customer usage patterns. It also answers customer queries about the performance of their services.

Figure 10.11 illustrates the main activities of the performance manager. These activities involve the performance data collection, management and analysis. The process also interacts with multiple service processes, peer SPs, and network management processes.

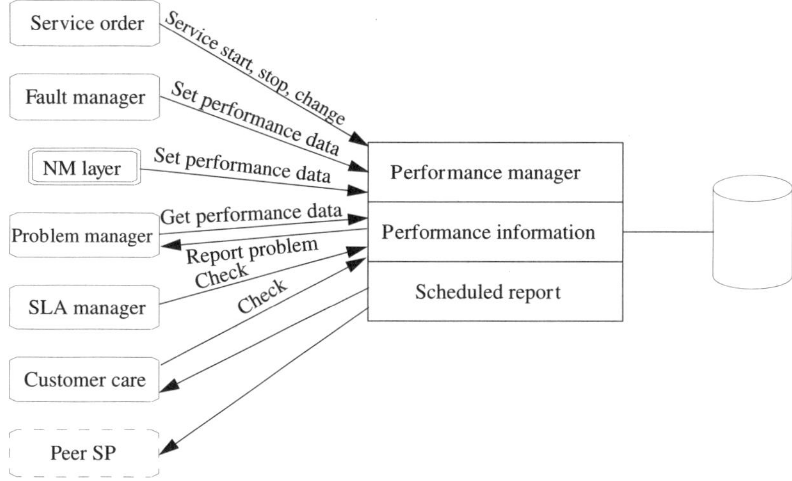

Figure 10.11 Activities of performance manager.

10.3.1.4 Problem Manager. The *problem manager,* normally invoked by trouble report process in the customer care management system, deals with customer service related problems. The process is entered when customers report service problems to customer care system. These problems include quality degradation, service interruptions, and service malfunctions. The process is responsible for locating and identifying the cause of the problems and repairing the faults. In many cases, operator involvements are required to identify and repair the problems. The process is responsible for generating SLA violation reports when some SLA terms are violated, such as the service downtime threshold being reached. Figure 10.12 depicts the workflow for the problem handling process.

In many cases, a single network fault may affect many services. It might be the case that a similar service problem has been reported by other customers and a trouble ticket has already been issued. It is important to associate this problem report with the existing trouble tickets so when the problem such as network fault is repaired, this customer is also notified.

One of the key tasks of this process is to exchange the problem reports with peer service providers and to establish the cause and location of the problem. This requires the interworking with the problem handling process of the peer service providers.

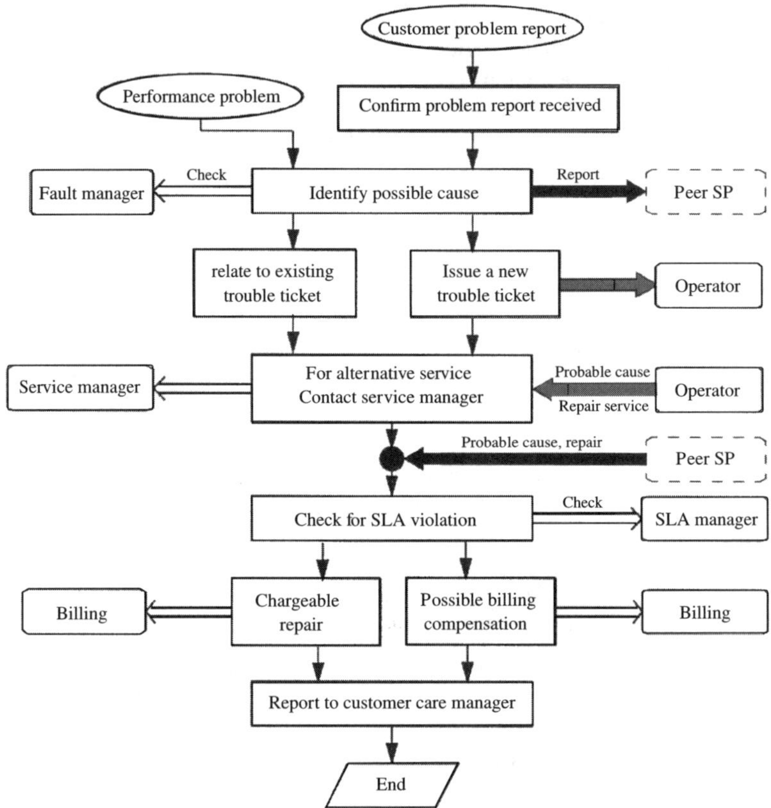

Figure 10.12 Problem manager process workflow.

Another task is to check whether and to what degree the SLA is violated due to this problem. This requires the interaction with the SLA manager. If an SLA violation is found, then the SLA violation report has to be generated by the SLA manager. Depending on the terms of the SLA, some billing adjustments may be needed. The problem handling process will interact with SLA manager and billing process for these purposes.

If there is no SLA violation and the repair of the problem is chargeable to the customer, then the problem manager will interact with the billing process to update the customer account.

In some cases, automatic problem repair may be supported (e.g., switching to an alternative service if the default one is affected by the problem). This may require the problem manager to directly interact with the service manager, configuration manager, and other network management processes to provision the alternative service.

10.3.1.5 Fault Manager. *Fault manager* is an internal process which generates service fault reports based on the network faults and alarms. A key capability of this process is to translate network faults and alarms into service faults. This process issues reports to the problem handling process and the customer care management system.

Similar to the service configuration process, the fault manager is a key process to support the integration between service management processes and network management processes. This process has the knowledge of both the service interaction model and the generic network model in order to support the processes and information flow-through between the two models. It is responsible for relating a network fault to a service defect which is defined as part of the service interaction model. An example of this mapping is that the network connection cut is translated to the faults in video connections, thus related to the customer's video-on-demand services. All other service management processes may choose to subscribe the service faults notification from the fault manager process. The customer care management can be used to notify service customers if required.

A fault manager also keeps all the service fault reports based on the network alarms and faults. These reports record the interactions with other service management processes such as SLA manager. It also automatically generates trouble tickets for certain service faults which need to be repaired by operators. Figure 10.13 illustrates the workflow for the fault manager process.

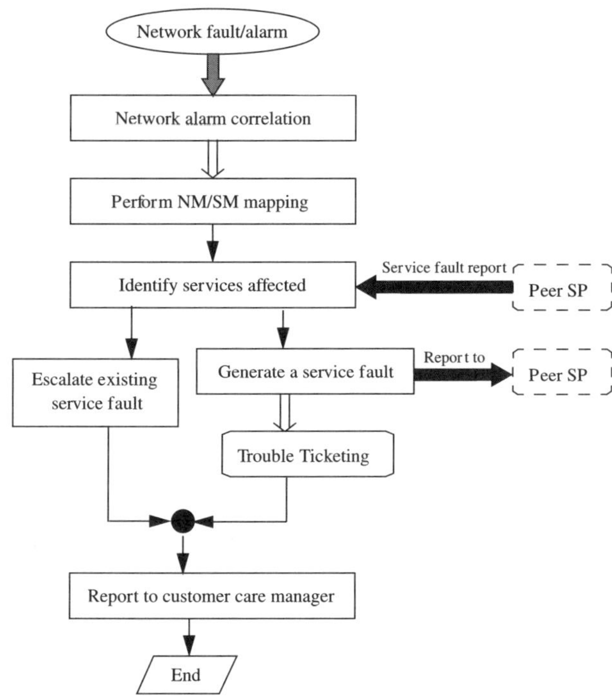

Figure 10.13 Fault manager process workflow.

10.3.1.6 Service Level Agreement Manager. The *SLA manager* is responsible for monitoring and managing quality of services as defined in the service parameters and SLA terms negotiated during the service ordering phase. It manages the SLA object lifecycle starting from the ordering process which creates SLA objects.

The SAL manager provides the quality of service status information for delivered customer services. These status information are sought by other service management processes including billing, performance and fault management. In addition to the static status of the SLA, the process is also responsible for generating scenarios for more proactive management. Examples of SLA scenario management includes:

- What SLAs will be violated by a particular maintenance activity?
- What SLAs are potentially going to be violated if network usage trends continue or if certain abnormal network behaviors occur?
- What SLAs are affected when information exchange with a peer service provider is interrupted?

The SLA manager also generates reports for the consumption by other service management processes, especially the customer care process:

- Service usage trends and patterns (e.g., customers who use feature X seem to want Y as well)
- Warnings to customers of potential violation of SLA (e.g., from performance degradation)
- SLA violations report

Figure 10.14 illustrates major activities supported by SLA manager.

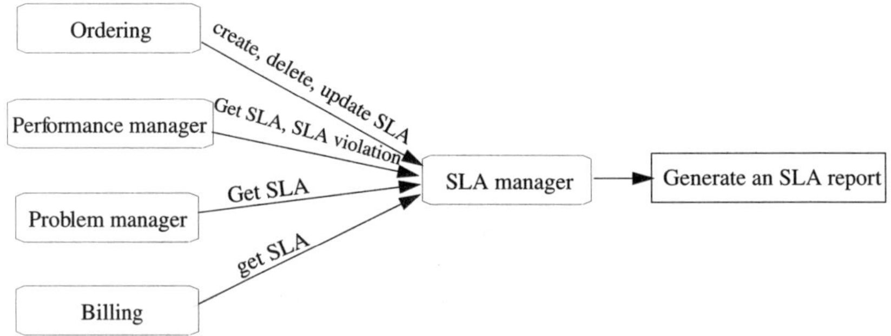

Figure 10.14 Major SLA manager activities.

10.3.1.7 Billing Process. The *billing process* is responsible for generating bills to customers and processing their payments. In addition, the billing process handles inquiries about call records, pre-bill information and other account information. Internally, billing process support interactions made by other service management processes for obtaining billing-related information. The billing process defines an integration point with the network management processes to obtain call records.

The billing process is also responsible for generating bills to other service providers for the services provided to them as part of the customer order. The process is also responsible for initiating the payment for the usage of services offered by other service providers.

Figure 10.15 illustrates the major activities of the billing process and its billing information management activities.

10.3.2 Service Processes Interaction

The service interaction model discussed in Chapter 9 defines the interactions among service management processes. These interactions are essential to achieving integrated service management. In the integrated architecture presented in Figure 10.5, each link between different functional blocks represents an interaction point between pair-wise processes.

Figure 10.16 illustrates in more detail the interaction points between different service management processes:

- The interaction between the service manager's ordering process and service configuration process. Service configuration maps service requests and service parameters to the network requests.
- The interaction between the service order process and the billing process. An example of this interaction is for the billing process to record the start and termination of a service.

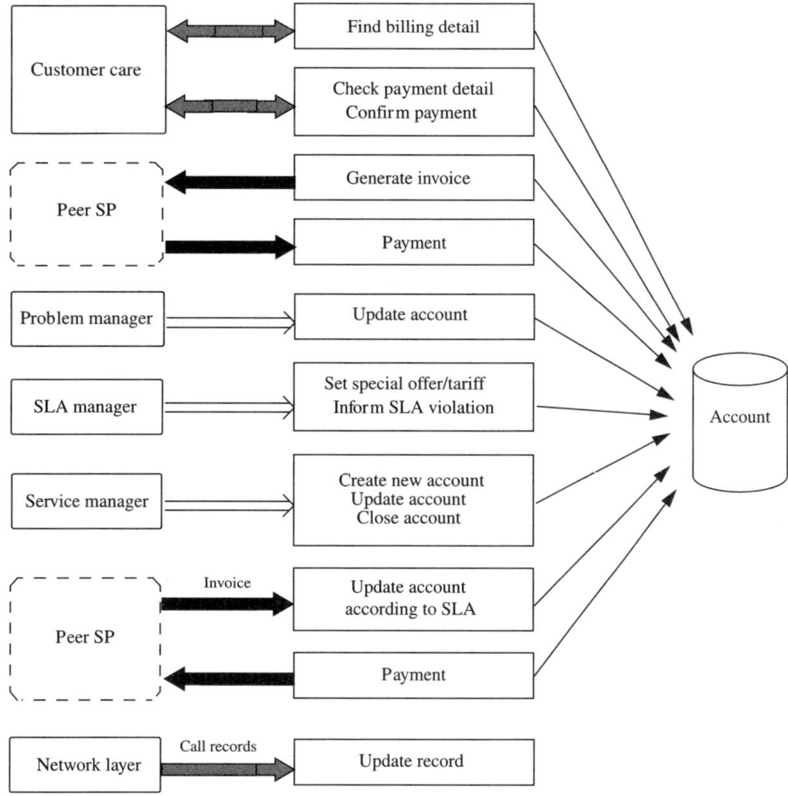

Figure 10.15 Billing information management activities.

- The interaction between the service order process and the SLA manager for SLA creation, deletion, update and negotiation.
- The interaction between the service order process and the performance manager. This interaction allows the automatic generation of performance monitoring criteria according to the SLA.
- The interaction between billing system and SLA manager. An example of this interaction is for billing system to retrieve the agreed SLA from the SLA manager. Some SLA may include a special rate negotiated during the service ordering phase, SLA compensation or repair charges.
- The interaction between performance and SLA managers. This interaction allows the performance manager to monitor SLA terms and to determine whether a performance degradation has caused the violation of the SLA.
- The interaction between the performance manager and the fault manager. This allows the establishment of whether the faults in network elements affect some services and their performance.
- The interaction between the performance manager and the problem manager. This allows problems to be reported and handled when poor performance is detected by the performance manager.
- The interaction between the fault manager and the internal trouble ticketing system. This interaction allows the fault manager to issue a trouble ticket automatically when a fault has be identified.
- The interaction between the fault manager and the problem manager. This allows the

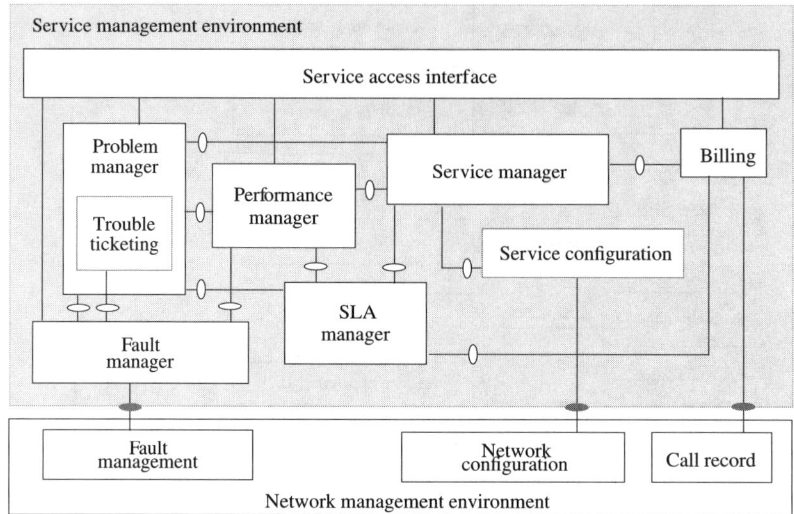

Figure 10.16 Interactions points between service management processes.

problem manager to check service fault records to determine the potential cause of the reported problem.

- The interaction between the problem manager and the service ordering process for the creation of an alternative service when repairing a problem for the existing service.
- The interaction between the problem manager and the SLA manager for obtaining the agreed SLA to determine the billing adjustment.

10.4 SERVICE INTEGRATION ENVIRONMENT

The service integration environment supports the integration between customer access environment and the service provider's environment defined in the integrated management architecture. It also supports service and network integration functions. In addition, the service integration environment supports the interworking with the peer SPs in a one-stop-shopping service provisioning scenario.

The service integration environment supports the following interface components:

- A service access interface which can be used by customer management layer and service management applications for issuing service management requests. These requests will invoke the service management process workflow. Potentially, the same interface can also be used by other service providers.
- A service management and network management integration layer which bridges two object models and associates service management requests to a set of network management requests.
- An SP-to-SP interworking component which supports OSS interconnections with peer service providers for service components.

10.4.1 Service Access Interface

Service access interface consists of a set of interfaces made available to customer management layer for requesting services. This service access interface is based on the functions supported by the service interaction model. These interfaces include:

- Pre-ordering
  —*pre-order request*—used by a customer to specify service requirements and to request a service offer based on these requirements.
  —*cancel pre-order request*—used by the customer to cancel the pre-order request.
- Ordering
  —*order request*—used by a customer to order services. The customer may order according to the offer resulting from pre-ordering interface, or service description resulting from the catalog search.
  —*cancel order request*—used to cancel the service order.
  —*change order request*—used to alter certain service characteristics and SLA details after an order has been made and before the ordered service is activated.
- Negotiation
  —*negotiation*—used by customers to negotiate with a service provider for the terms and conditions of a service during service pre-order or ordering stage. The scope of negotiation includes the service features, special offers, SLA, cost, etc.
- Delete service
  —*delete service request*—used to terminate an existing service.
- Problem reporting
  —*problem report*—used to report problems to the service provider with a reference to the service exhibiting the problems.
- Account payment
  —*payment*—used to make a payment to a customer account with a service provider.
- Enquiries
  —*billing enquiry*—used to check the balance of an account.
  —*SLA enquiry*—used to retrieve details of the SLA for a particular service or to retrieve all SLAs with a service provider.
  —*fault report*—used to retrieve a fault report which may affect the customer's services.
  —*performance enquiry*—used to obtain performance statistics on a service or specified type of services.
- Processing status enquiry
  —*query pre-order request*—used to obtain current processing status of a pre-order request.
  —*query order request*—used to obtain current processing status of a order request.
  —*query negotiation request*—used to obtain current processing status of a negotiation request.
  —*query trouble report processing status*—used to obtain current processing status of a trouble report.

10.4.2 Interworking between Service Providers

An important function of the service interaction model is to support the interworking between service providers. This function forms part of the service management process automation. The SP-to-SP interworking enables the peer service providers to cooperate in an end-to-end service provisioning and management scenario, where multiple service providers are involved in a single customer request for services. This function is also important to support one-stop-shopping, so that the customers' requests are managed by one single service provider of their choice.

Figure 10.17 illustrates the SP-to-SP interworking scenario in which a service provider may play two different roles. In scenario (a), the service provider A plays a customer role. It requires services from other service providers B and C. In scenario (b), SP A plays a supplier role by providing services to other service providers.

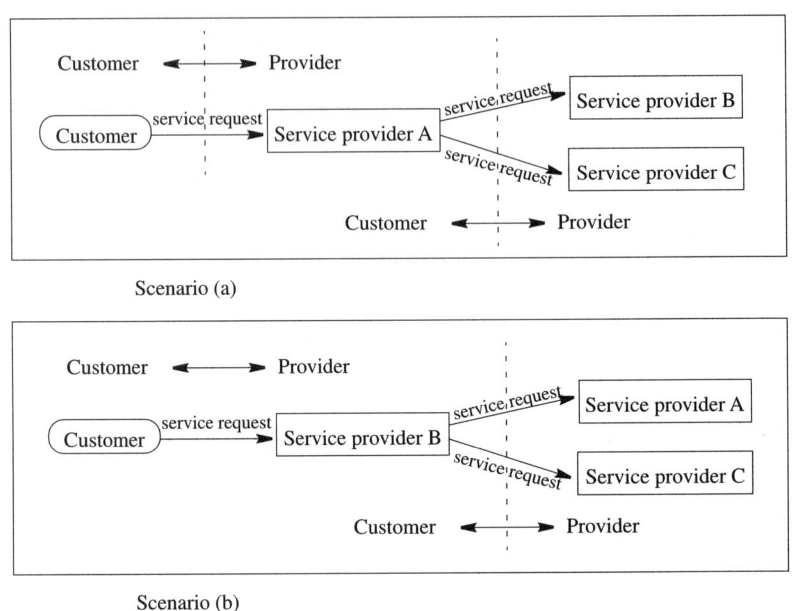

Figure 10.17 SP-to-SP interworking scenarios.

When playing a customer role on behalf of the end customer requests, a service provider is able to:

- Issue service requests, including service order requests, status checking requests, etc. Different from end customer requests using interactive customer care interface to access the supplier's interface, most SP-to-SP interworking activities need to be automatically performed in a predefined manner.
- Receive responses from peer service providers including service order responses, problem handling responses, etc. These responses need to be dispatched to different service management processes for them to take appropriate actions.

When playing a supplier role, a service provider receives service management requests, processes them and responds to them in a same way as it handles end-customer initiated service requests.

There are two architecture approaches to support SP-to-SP interworking. The first approach is a tightly coupled approach such that the interworking SPs both adopt the same business process model and business object model. The second approach is for interworking SPs to rely on mediation devices, commonly referred to as the OSS interconnection gateways, to achieve process and information exchange.

Using the tightly coupled approach requires a tighter integration between the two service management environments. A generic service interaction model and its common processes need to be supported by both interworking service providers. The software integration can be made easier if both providers support the model using same technology and interface definitions.

This approach requires the following interworking functions to be supported by the service management processes:

- *Service manager*—it is required to understand the scope of services provided by other interworking service providers. These services or components form integral part of the service and product offers managed by the service manager. This knowledge includes the service features, quality of service parameters, service cost, and the overall SLA framework. The service manager is able to order service components from the interworking service providers and handle exceptions when required components are not available or cannot be delivered within the required time frame and SLA terms.
- *Problem manager*—it is required that the problem manager be able to invoke processes to the interworking service providers whose services are part of the overall service offer to the end customers. The functions allow the problem manager to enquire, locate, report, and repair the cause of the problems.
- *Performance manager*—it is required to receive performance reports for interworking service providers for the subordered service components.
- *Service fault manager*—it is required to receive fault reports for the subordered service components affected by the faults.
- *Billing process*—it is required to query account information and settle the accounts with the interworking service providers for the subordered service components.

Although the SLA manager is required to incorporate SLAs for subordered service components as part of the overall SLA to the end customers, it is not necessary for the SLA manager to interact directly with other service providers. Instead, it manages those sub-SLAs on behalf of other service management processes including service manager, problem manager and billing manager. These processes interwork with their peer processes to negotiate SLA terms, report SLA violations and adjust billing records.

In a tightly coupled integration environment, these functions can be provided as the extension to the existing functions supporting the customer-to-provider relationship. For example, the core of the customer care system which interacts with the service management processes can be extended to act as the gateway to interact with the peer SPs processes. Further, this interaction layer may well be a logical component in an actual software design. For example, if CORBA is used by both interworking service providers as the service management platforms, interactions between these two service providers become interactions between cooperative CORBA systems. SP-to-SP service requests are invoked to the same objects representing ordering, problem handling, performance processes, but living in different service management domains.

Figure 10.18 illustrates the process interworking between two tightly coupled service providers.

The second interworking approach is based on the principle of satisfying the minimum interworking requirements. In this approach, two interworking SPs are not required to have common business processes and information models for their service management functions and interfaces. Their service management environments may not interoperate at the process level, nor do they interoperate at the technology level, as the service management environments may be supported by two different and noninteroperable technologies.

This approach requires a minimum set of interconnection gateways to be defined and these gateways operate as mediation devices between two service management domains. The example of this approach includes the so-called OSS interconnection gateways adopted by the industry to mediate the wholesale–retail relationships.

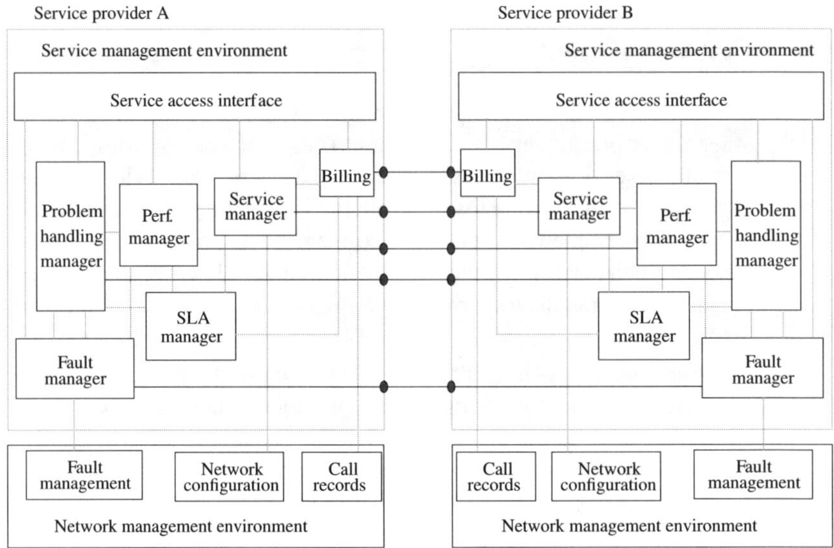

Figure 10.18 SP-to-SP process interworking.

The interconnection gateway approach defines a set of business processes common to every service provider. Each gateway defines and mediates a specific business process. The gateway supports different protocol definitions at two end points. The service providers are required to integrate these end points with their own service management environment. Figure 10.19 depicts the overall approach of the interconnection gateways.

In Figure 10.19, each gateway mediates a different process and supports different protocols at both ends. The dotted ellipse indicates the integration between the gateway and the SP's OSS systems. The processes defined in the service interaction model can be used to model interconnection gateways. However, it is not necessary to support all processes, nor is it necessary to use the same software architecture to support them.

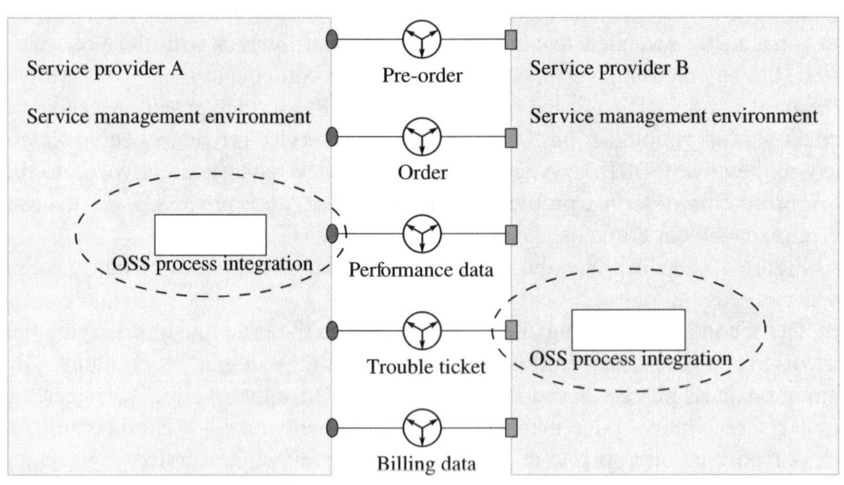

Figure 10.19 OSS interconnection gateways.

10.4.3 Service and Network Integration Layer

The service and network integration layer plays an important role in the entire integrated management architecture. This layer is responsible for bridging the differences between the service management environment and the network management environment. In this architecture, this layer acts like a vertical mediation device between the service interaction model and the generic network model.

There are two aspects of the integration functions supported by this layer: the integration between the different object models, and the technology integration strategies.

The object models representing service management processes and network management processes are quite different. Service object model represents more dynamic complex services and the network object model represents relatively stable and less complex but finer grain network objects. The service and network integration layer needs to understand both these object models and is responsible for the integration between them. Figure 10.20 illustrates the main function of this layer.

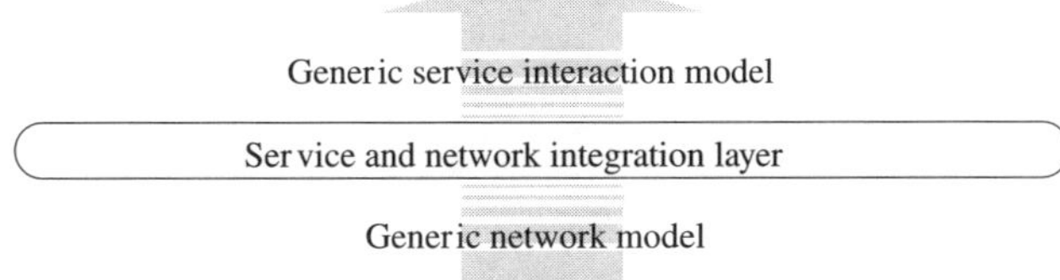

Figure 10.20 Mediation of different object models.

As discussed in the previous chapter, the generic network model represents the end-to-end network management functions aimed at supporting service management processes. The service interaction model, on the other hand, is designed to support the interaction and interworking between service management processes. The service interaction model is thus directly supported by the generic network model to access network management functions. Following the concepts of abstraction and specialization, the service interaction model is treated as a higher level abstraction of generic network model and this integration layer provides the architectural support for this abstraction.

Two service management processes discussed in this chapter are example components of this layer: the *configuration* process and *fault management* process. The configuration process converts a customer order provisioning task into a set of network resource configuration tasks. In order to perform this conversion, the process supports the service interaction model by using the topology and connection management functions in the generic network model. It maintains the relationship between customer services and network topology and connections. This frees the customer care system and service management system from having to know the precise and detailed information about the network infrastructure. For example, this layer handles the conversion between a video conference link to a broadband connection which involves ATM network and access network. This allows the customer care system to link the network connection status to the existing video conference service instances and their customers.

The fault manager converts network alarms and events in the generic network model into the fault reports in the service interaction model, and it manages the relationship between the network faults and services. This capability contributes significantly to the service management system and the support for a better customer care management system.

The second aspect of the service management and network management integration deals with the technology integration strategies. It is quite common that different management technologies and platform products are used to manage services and networks. Therefore, the service and network integration layer is also responsible for the integration between different management technologies. For example, when CORBA is used to support service management functions and OSI is used to support network management functions, then this integration layer will be responsible for implementing CORBA/OSI integration strategies discussed in Chapter 8. This is achieved by using generic network model to act as CORBA/OSI abstract integration gateway to provide a higher level interface functions to support service management systems.

10.5 NETWORK MANAGEMENT ENVIRONMENT

The network management environment supports integrated network management functions. This environment contains two major components, as illustrated in Figure 10.21:

- A generic network model which provides the service management layer with a generic network management interface. This model offers the higher level network management functions to support the integration with service management.
- A set of integration points (integrators) to specific network management systems (NMS) or network element management systems (NEMS).

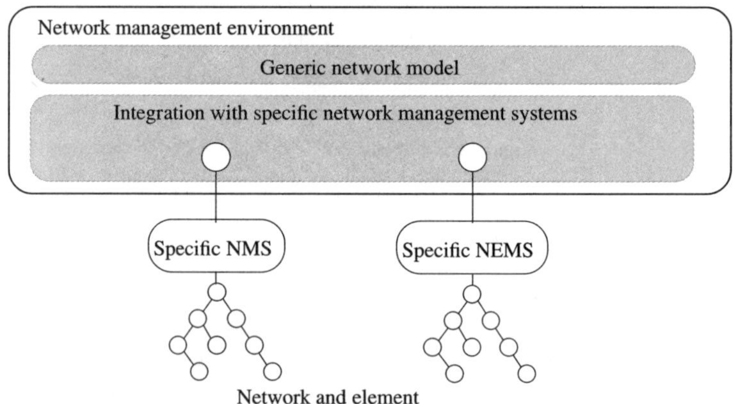

Figure 10.21 Network management environment.

10.5.1 Generic Network Model

Generic network model presents a generic network view to the service layer. It contains a number of functional submodels such as topology model and connection model. The basic topology model represents the physical topology of the managed network, the network equipment (switches, routers, multiplexers) and their connectivity. The global logical network view can then be constructed based on these elements, for example, SDH/SONET rings, administrative domains or ATM virtual LANs, etc. The connection model represents the service-oriented views of logical connections created dynamically in this physical network.

The generic network model serves two purposes in the integrated management architecture. One is to support the service management layer by only exposing the service access points and encapsulating the complexity of the physical network structures. The other is to provide a layer of integration between different network technologies.

A network management system based on the generic network model supports network management applications which are developed for network operators. These applications perform the network administration, operation, management, and maintenance tasks. The same network management system also supports service management systems to perform day-to-day service management activities.

10.5.2 Integrators

The integrators in the network management environment are set of integration points between the generic network model and the underlying specific network management systems. Each integrator supports a set of generic interfaces defined in the generic network model and maps these interfaces into the underlying network technology it supports.

An integrator provides the following abstract views:

- *The topological view of the managed network.* This view describes the number of subnetworks contained in the managed network and the way they are connected. This view serves two purposes. It provides the application with a simplified network topology and services and it represents the managed network's internal structure. This internal structure is used by the internal algorithm to determine how a management request can be delegated to its subnetworks.
- *The network service view.* The network service view represents the ways the network can be used to transfer data from one place to another. The network service view includes the external network access points (terminations) and the types of connections that can be established between these access points. The network service view is made available by a network to its users, such as service management layer, and its super-network.

Hence, each network integrator performs two major tasks. One is to maintain the rules of abstraction and specialization between specific network management system and the generic network management system represented by the generic network model. The other is to translate a generic network model management request into a set of specific management requests supported by the underlying specific network management systems.

In this section, we discuss this specialization and abstraction in more detail. We use managed networks which support CMIP as an example.

Let's assume that a logical network, `SuperNet`, modeled in generic network model represents a physical network that is partitioned into two subnetworks, `SubNet1` and `SubNet2`, as illustrated in Figure 10.22.

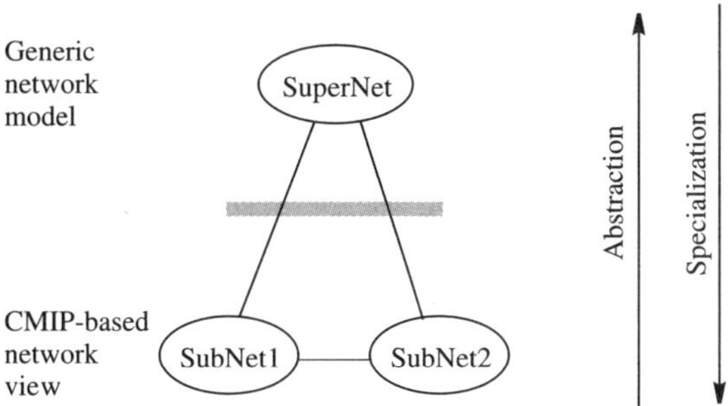

Figure 10.22 Generic network model network with CMIP subnetworks.

Both subnetworks have a CMIP-based management interface. During the abstraction, the first activity is to identify the information provided by each of the subnetworks. This includes:

- The external view of each subnetwork, such as
  —a set of terminations which delineates the subnetwork
  —the types of connections supported by each subnetwork
- The topological link between subnetworks

The information of each `SubNet` is provided to the `SuperNet` so that the `SuperNet` understands its own topological structure.

Each of the two subnetworks may contain a set of further interlinked subnetworks. These information are internal to each `SubNet` as they represent their own topological structure. The management within the subnetworks are performed by CMIP-based manager.

During the abstraction stage, a logical network, `SuperNet`, is created in the generic network model. This network has an abstract view of its subnetworks. Each logical terminations in the `SuperNet` is an abstract view of one or a group of terminations in one subnetwork. The information collected during the abstraction activities includes:

- relationships between the logical terminations in `SuperNet` and the terminations in the subnetworks `SubNet1` and `SubNet2`.
- transport layer of each subnetworks. This information will determine the transport layer of the logical network.
- topological links between subnetworks. This information will determine the logical connectivity of the subnetworks.

Since the super-network, `SuperNet`, represents an abstraction network and encapsulates the detailed topology of the two subnetworks, the clients of the super network only see the external view of the super network and can only request connections over the super network. This abstraction view of `SuperNet` and its subnetworks is illustrated in Figure 10.23.

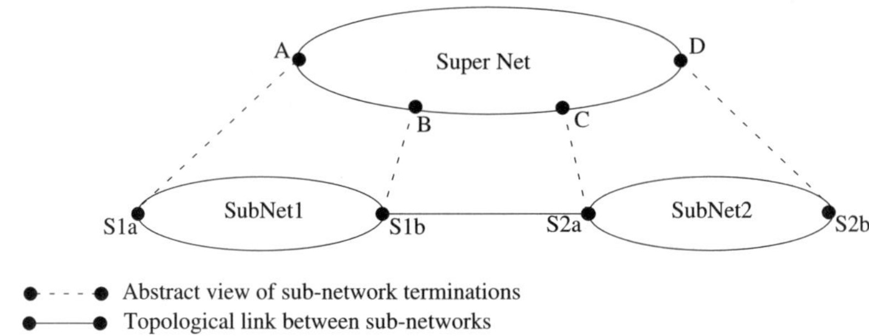

Figure 10.23 Abstract view of the managed network.

The information obtained during abstraction process is maintained by the super-network. This information is used for the super-network to understand the information received from the underlying subnetworks, such as alarms, events, etc. Based on the abstraction rules, an alarm generated by an element in a subnetwork is mapped onto the logical network which represents the abstract view of the managed network in generic network model. This information enables the generic network model to identify the affected logical connections and prevents applications from creating new connections using the faulty network element.

The information modeling process also defines the specialization rules which are used by the super-network when invoking management functions to the subnetworks. For instance, when applications require `SuperNet` to create a connection between two logical terminations A and B, this creation request is mapped onto a set of CMIS requests to the subnetworks following the abstraction and specialization rules. If A and B are abstract views of terminations S1a and S1b respectively, in the same subnetwork (e.g., SubNet1), then the creation request is translated into a set of CMIP requests to SubNet1. Among the requests is a cmis_create to create a connection between S1a and S1b. The CMIP agent is responsible to perform the creation in the MIB and on the switches involved. Such a connection creation is illustrated in Figure 10.24.

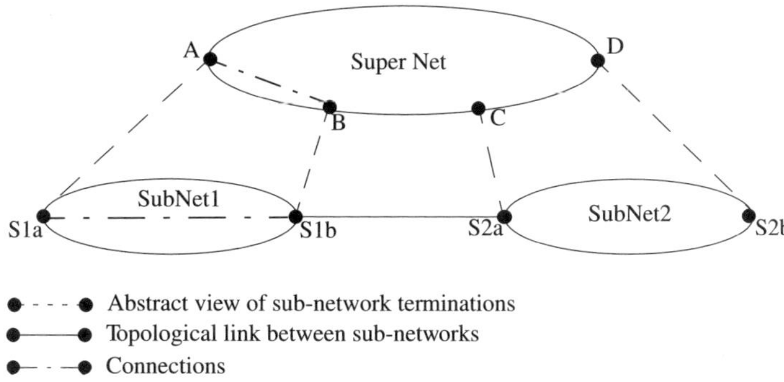

Figure 10.24 CMIP specialization (1).

If the connection creation is between A and D, and A is an abstract view of S1a in SubNet1, and D is an abstract view of S2b in SubNet2, and both subnetworks are linked by a topological link between S1b (in SubNet1) and S2a (in SubNet2), then the connection creation request is translated into a set of CMIP requests which includes two cmis_create operations as:

- Creation of a subnetwork connection from S1a to S1b in SubNet1
- Creation of a subnetwork connection from S2a to S2b in SubNet2

A link connection over the topological link between S1b in SubNet1 and S2a in SubNet2 is created within the SuperNet because it is over a topological link managed by the SuperNet. The CMIP management systems for both subnetworks have no knowledge of the link at all.

The required connection in the super network is an abstract view of the connection across two subnetworks. This is illustrated in Figure 10.25.

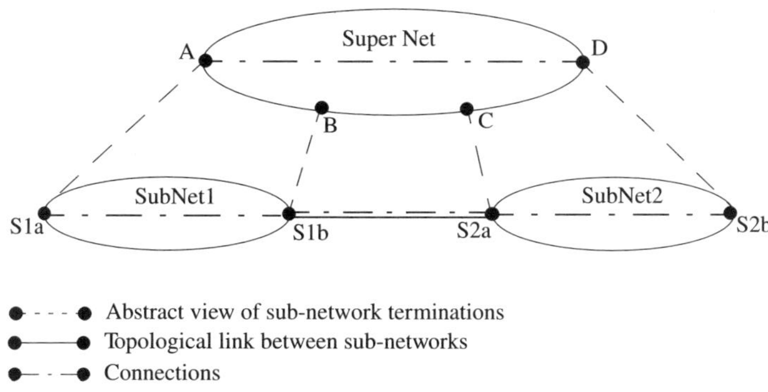

Figure 10.25 CMIP specialization (2).

10.6 SUMMARY

The integrated architecture presented in this chapter deals with the integration of two different process and information models: the service interaction model and the generic network model. The integration of these two models provides a platform to combine different management subsystems and components in a single architecture.

The management subsystems and components discussed in this chapter consists of:

- Customer care and management system
- Service management OSS components and processes
- SP-to-SP interconnection gateways
- Network and element management systems

The service interaction model and the generic network model provide a framework for integrating, (rather than reengineering) these components to achieve the business objectives.

This chapter also discussed multiple types of integrations:

- Customer care and management to service management integration
- Service management business process interaction
- SP-to-SP service management interworking through interconnection gateways
- Service management to network management integration through generic network model
- Generic network model integration to specific network technology through inetgrators

Different integration approaches discussed in Chapter 8 were employed in this integrated architecture. The abstract mapping approach is primarily used for semantic abstraction of management functions. Although the primary example of this approach is given in the CORBA/OSI integration context, the approach is not limited to the integration of these two technologies alone. For instance, in the service management environment where a single technology such as CORBA is used, it is still the essential function of the integrated architecture to generalize management functions to present more abstract view of services and service management capabilities. The generic network view needs to be generalized to support the services and service management processes and business objects. These processes and business objects need to be generalized to support customer care functions and customer management processes.

The architecture also support horizontal interactions and interworking between multiple management processes. It does that by supporting business process workflow and their integration with the business objects. The application interworking approach is also applied to the SP-to-SP interworking scenario to support more transparent customer services.

In the next chapter, we focus on an exercise to map this architecture to software solutions. We examine the software technologies that implement the business processes and integration. We apply the processes and principles of building distributed systems and integration strategies to form integrated management solutions.

11

Building Integrated Management Solutions

So far, we have discussed the business requirements for integrated management solutions and explored different integration approaches. We have also defined the generic object models that support the service process interactions and integration between service and network management systems. We have further defined an integration architecture that supports these models by defining business processes and customer services and the relationship between them. In this chapter, we present a software architecture and technology framework to implement these business processes and interactions.

11.1 INTEGRATED SCENARIO

In this section, we discuss the architecture and high level design of the integrated management solutions. Figure 11.1 depicts the typical interactions in a service provider's management environment. This interaction scenario shows the following basic components in each of the service provider environments:

- *Service access interface*—the access point for all the service management functions.
- *Service manager*—performs service ordering, configuration and management functions with the support of the back-end workflow processes.
- *Problem manager*—handles problem report and management with the support of back-end, trouble-ticketing management.
- *Billing process*—handles all billing-related activities.
- *Performance manager*—monitors the service performance and identifies possible SLA violations.
- *SLA manager*—manages the impact of overall service performance on customer service level agreements. The process uses SLA as the main interaction point between different service management processes.
- *Fault manager*—maps network fault into service fault.
- *Network management environment*—presents to the service layer an abstract view of the network resources and network management functions.
- *Managed network*—provides the network resource support for the services provided.

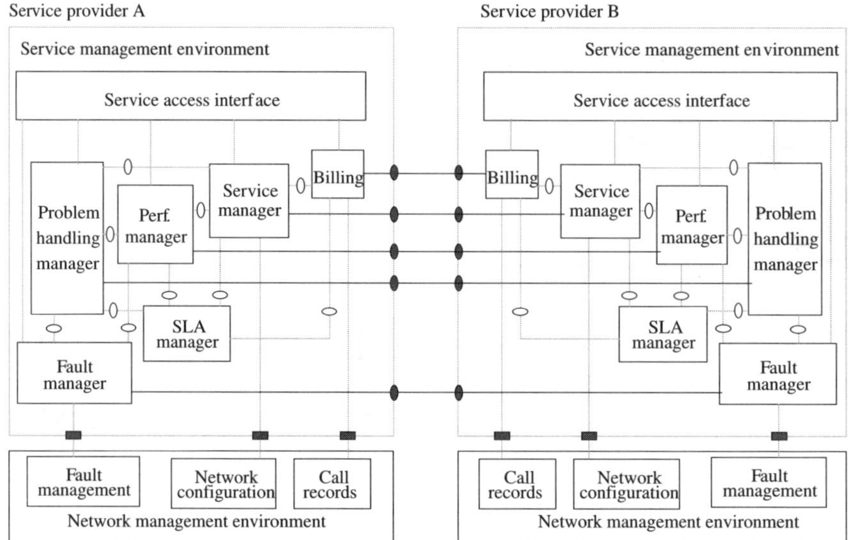

Figure 11.1 Business processes integration scenario.

There are three types of interaction points as depicted in the Figure 11.1. These interaction points are:

- The interactions between different service management processes within a single service provider's domain.

  Support for these interactions enables a common interface to both service management applications and customers to access integrated service management functions. In this architecture, the implementation of the service interaction model provides the support for this type of integration.
- The interactions of service management functions between different service providers. In a global deregulated environment, these interactions are required to achieve *one-stop shopping* for end customers. Agreed common interfaces between different service providers are required in order to support this service level interoperability.
- The interactions between service and network management processes. Support for these interactions enables the management processes and management information to flow through the service and network management boundaries. The implementation of the generic network model in this architecture provides support for this integration.

The software architecture described in this chapter defines a management platform which helps developers use appropriate technologies to support business processes. It is focused on customer solutions, not technologies.

We define an integrated management platform which has the following major components:

- *Customer services environment*—This environment provides a well-integrated customer interface to the service provisioning and management environment. It is a cross platform, lightweight user environment which offers service access functions for operators and end users.

- *Service management environment*—This environment implements business application semantics, customer and service management functions, workflow for business processes. This environment also provides distributed infrastructure support.
- *Network environment*—This environment provides a downward integration between the service management environment and the network environment.

The systems architecture is designed to satisfy the following business objectives:

- To use the service interaction model and its management processes as the underlying model to define management functions.
- To use the service interaction model as an integration platform for multiple business processes and to support integration between different service providers.
- To implement a set of common business scenarios. The technologies used in the architecture are the enabling technologies which provide the functional capability, flexibility, and reliability the platform requires. It is expected the architecture will survive the impact of technology evolution and migration.
- To use the generic network model as the integration platform for service and network management integration.

The maturity in the telecommunication management industry is evident by the change in mentality from *seeking the best management technology* to the *coexistence of multiple management technologies.* The key issue now is how to obtain benefits from the integration of different technologies in creating management solutions. The ability to reconcile different management models is key to achieving successful integration. The software architecture discussed in this chapter focuses on the integration between these models.

11.2 SOFTWARE ARCHITECTURE

Figure 11.2 depicts the high level software architecture for the integrated management functions.

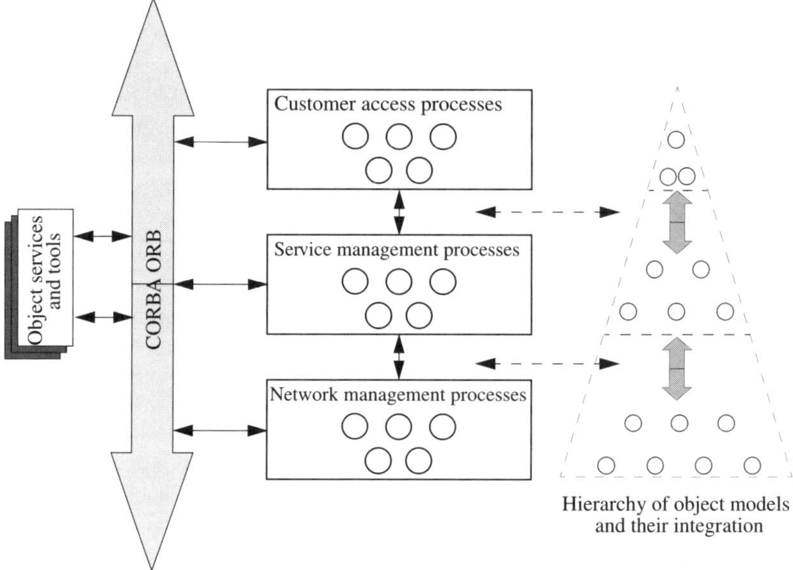

Figure 11.2 High-level software architecture for integrated management.

The software architecture identifies three major functional components. The customer access and customer care component provides functional support for end customers to interact with the system. It also provides user interfaces for business processes supported by the service management components. In the software architecture, this component needs to be computer platform independent and be able to support different user-interface technologies. The Java/Web technology is ideal for customer access component development for Web and Internet-based access. Other technology includes IVR and its integration with a call center system. The customer access component has an interface in the service provider's customer service representative side for the customer care and management. It also provides an interface at customers side.

The service management component supports business processes discussed in the Chapter 10 and the service interaction model discussed in the Chapter 9. The service management component requires the integration between distributed object-oriented technology and process flow to support its management needs. CORBA is selected as the distributed platform in this architecture. The service management components are located in the service provider's service centers. CORBA platform allows different processes to be distributed in different locations.

Frequently, many service providers have already developed network management systems for its telecommunication networks. These systems may be OSI CMIP-based or SNMP-based implementation of network/element functions. They may also be systems using proprietary management protocol and technology. These components need to be integrated. In this architecture, we focus on the integration with OSI CMIP-based network management systems. The abstract object definition method discussed in Chapter 8 is used as the integration approach between the CORBA based service management component and OSI-based network management component.

The architecture identifies three major components corresponding to the three major management functions. However, we don't suggest three different computation environments or communication stacks. All three components are integrated by a common CORBA computation bus as illustrated in Figure 11.2. The horizontal arrows between each component and ORB indicate this shared computing environment. The vertical arrows between three components indicate the object abstraction and specialization. In an architectural sense, a component supports the functions of the next level up by implementing a more specialized and detailed object model, and a component abstracts the function of the next level down by implementing a more general object model. The triangle on the right-hand side indicates this relationship of object specialization and abstraction. In a computational sense, these arrows represent the object mapping and technology integration points. For instance, the arrow moving from the customer access component to the service management component represents the technology integration between Java/Web and CORBA. The arrow from service management component to network management component indicates the technology integration between CORBA and OSI.

From this high-level architecture, we move to present different component architectures with more details. We focus on the following architecture components:

- Process flow and process control
- Customer access system
- Service management system
- Network management system

The rest of this chapter discusses these different components in detail.

11.3 TECHNOLOGY FRAMEWORKS

To support the implementation of the integrated architecture, the following three technologies play the key role.

- Technology to support distributed objects
- Technology to support object integration
- Technology to support process flow.

Figure 11.3 illustrates the combination of three technologies in an integrated architecture. The architecture supports the business process flow and their integration with business objects. These business objects interact and communicate with each other through a distributed CORBA integration platform. The architecture also supports the integration between these business objects and other computational objects (OSS and managed network objects) which perform OSS functions and specific network management functions implemented in specific technologies.

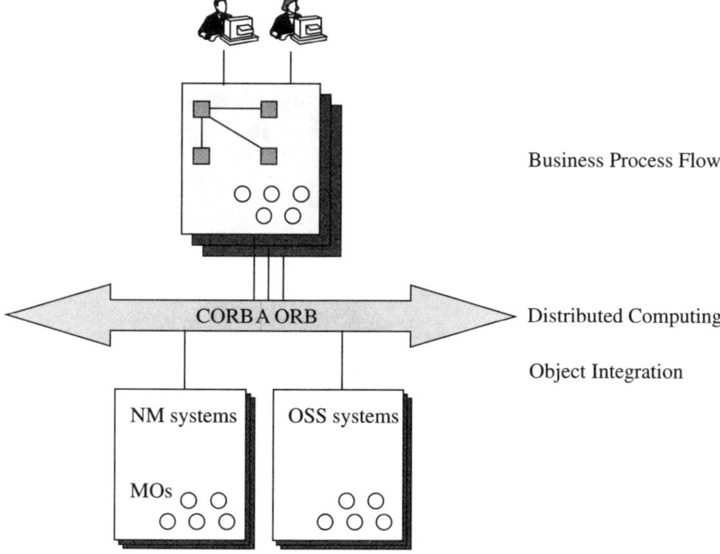

Figure 11.3 Technology support for integrated architecture.

The distributed object technology such as CORBA plays a key supporting role in a distributed computing environment. This environment provides most of the plumbing functions which allow objects at all different levels to interact and communicate in a unified manner. Chapters 5 and 6 have already demonstrated the usefulness of the technology.

Another key element of the integrated architecture is supports object integration at different levels. In an integrated management system, it is very likely that different components are implemented by using different technologies, such as CORBA, Java, and OSI. The integrated architecture must offer support for the integration of these different object models. Chapter 8 has presented various approaches of object integration for various purposes. These approaches can be used to achieve two different integrations required here:

- The integration between service interaction model and generic network model. This is an integration of two CORBA object models at different abstraction levels.

- The integration between generic network model and the underlying managed object model. This is an integration of the CORBA object model and the GDMO object model.

In this section, we discuss another key element of the integrated management architecture: the support for business process flow. The technology we target here is the transactional workflow [44]. The business process integration architecture requires a combination of functionality between workflow management technology and transaction processing technology. The workflow technology offers the functionality of process definition and control, and the transaction processing technology offers the functionality of data consistency, concurrency, and support for scalability.

A workflow management system is a system that completely defines, manages, and executes workflow processes through the execution of software whose order of execution is driven by a computer representation of the workflow process logic. It is a technology supporting the reengineering of business and information processes. The workflow technology provides support for:

- Defining workflow, in other words, describing those aspects of a process that are relevant to controlling and coordinating the execution of its tasks
- Managing workflow, for example, starting tasks according to the order and condition specified in a workflow, keeping track of all the tasks, and keeping a record of workflow processes
- Executing automatic tasks when required
- Providing for fast design and implementation of the processes as business needs and information systems change

Each management process described in the previous chapter can be regarded as a workflow process and can be managed by a workflow management platform.

Major business requirements for transactional workflow fall into the following categories:

- *Enterprise requirements*—Coordination of activities at an enterprise level requires support for defining and implementing task dependencies, business rules, and temporal information across different business domains.
- *Distribution requirements*—The move to distributed systems requires that workflow management systems be integrated into distributed, heterogeneous environments.
- *Reliability requirements*—When workflow management systems are applied to more critical areas of business operations, the ability to guarantee the integrity and reliability of the processes and data becomes significantly more crucial to the success of such systems.

All these requirements are targeted by the integrated management architecture defined here.

On the one hand, distributed transaction processing technology complements the workflow management in various ways:

- It brings an enhanced reliability, correctness, and consistency into the workflow management.
- It allows complex task dependencies to be specified. For example:
  —*sequential tasks:* one task starts after the successful completion of another task.

—*parallel tasks:* a number of tasks can be started at the same time and performed concurrently. A follow-up task is started only after the successful completion of all the parallel tasks.
—*alternative tasks:* a number of tasks are started at the same time and performed concurrently. Only one of the tasks is allowed to complete. All other tasks must roll back to a predefined stage. Preference can be specified among the tasks.
—*conditional tasks:* the task starting next depends on the previous task result.

- It involves some long-lived transactions, which means that resources cannot be locked for the whole life time of such transactions. The traditional ACID properties of transaction processing need to be extended to include:
  —*compensational transactions:* these are executed when a rollback is required for a committed transaction. In a normal TP environment, a transaction cannot be rolled back after it has been committed. However, in many cases, a compensation can be defined and applied to a committed transaction to reverse the effect of the transaction.
  —*customized transaction specifications:* these allow users to define their own transactions, and the rollback and compensation behaviors of transactions.

On the other hand, the introduction of transaction functionality in workflow applications requires some extra support in addition to the existing functionality offered by standard transaction technology. In particular, the following features are required:

- *nested transactions*—support multilevel transactions that allow committed tasks to be rolled back if required
- *rollback behavior definition*—allows customized definition of rollback semantics
- *compensation actions*—support multilevel transactions that allow committed tasks to be rolled back if required—an alternative mechanism to nested transactions
- *specification of semantic dependencies*—allow extra semantic constraints to be specified and enforced for a workflow
- *object orientation*—utilizes the functionality provided by distributed object technology and in particular the transaction processing support for workflow systems.

The combination of these two technologies makes the transaction workflow management an important technology choice for implementing the business process architecture.

Transactional workflow management can be integrated with the available middleware technologies such as CORBA. CORBA provides the basic communication capability in a distributed environment, supports standard interfaces to applications and supports for integration of multiple communication protocols. CORBA also provides good support for the business objects essential to the workflow environment. Finally, the interoperability between heterogeneous tasks can also be well supported by CORBA.

Figure 11.4 depicts a workflow design tool which generates application templates according to the business processes specified by the business process modeler. Figure 11.5 depicts a scenario of telephone service provisioning represented as a workflow process. This scenario is used to demonstrate different relationships between workflow tasks. It does not follow exactly the service ordering process defined in rest part of this book.

This service provisioning workflow consists of the following tasks:

- Information collection including:
  —collecting information from the customer
  —verifying the accuracy of the information collected
  —creating a corresponding service order record

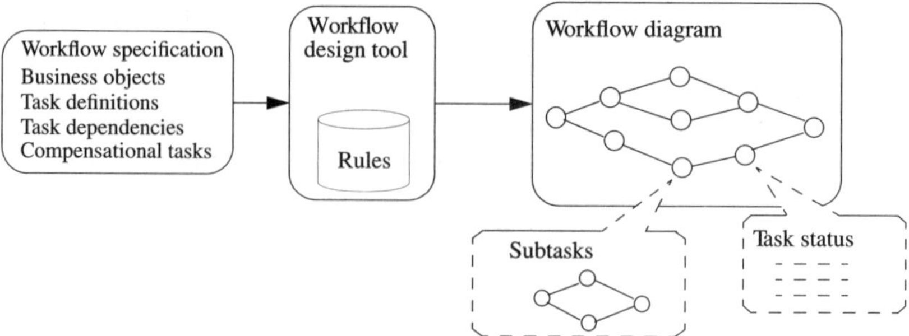

Figure 11.4 Workflow design tool.

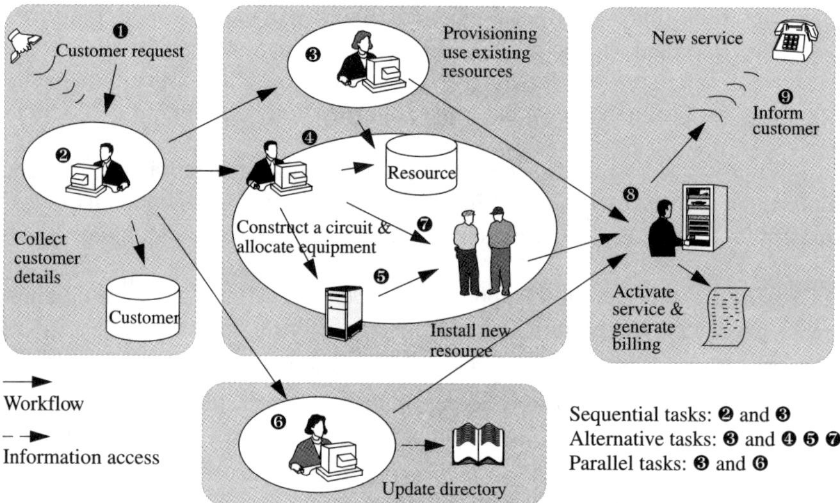

Figure 11.5 A service management workflow.

- Service provisioning, which involves constructing a circuit from a customer location to the appropriate telephone switch and allocating equipment to connect the circuit. This activity includes:
  —locating existing resources, such as lines and slots in switches
  —installing new resources if no existing resource is available
- Updating the telephone directory.
- Activating and testing the new service, informing customer that the new service is available and generating bills.

The provisioning workflow exhibits the following features:

- *Task dependencies*—Examples of task dependencies in the workflow are:
  —*sequential tasks:* a service order is created after the customer information is collected.
  —*parallel tasks:* service provisioning activities and changing directory action can be started at the same time. When both tasks are completed, next task can be started.
  —*alternative choices:* two service provisioning activities can be started at the same time, but only one is allowed to complete. If an existing resource is available and the provisioning is successful, this resource will be used and the other provisioning task is rolled back. The rollback semantics can be user defined.

- *Rollback behavior*—among two service provisioning activities, only one task can complete. If the existing resource is available for the service, the task which intends to use the existing resource should commit. Concurrently, the task that installs new resources should roll back. However, the rollback behavior of the second task can be defined as to complete the activity and record the newly created resources in the resource database for later use.

Transactional semantics is required to guarantee the correctness of such workflows. Figure 11.6 illustrates such semantics in the example service provisioning workflow.

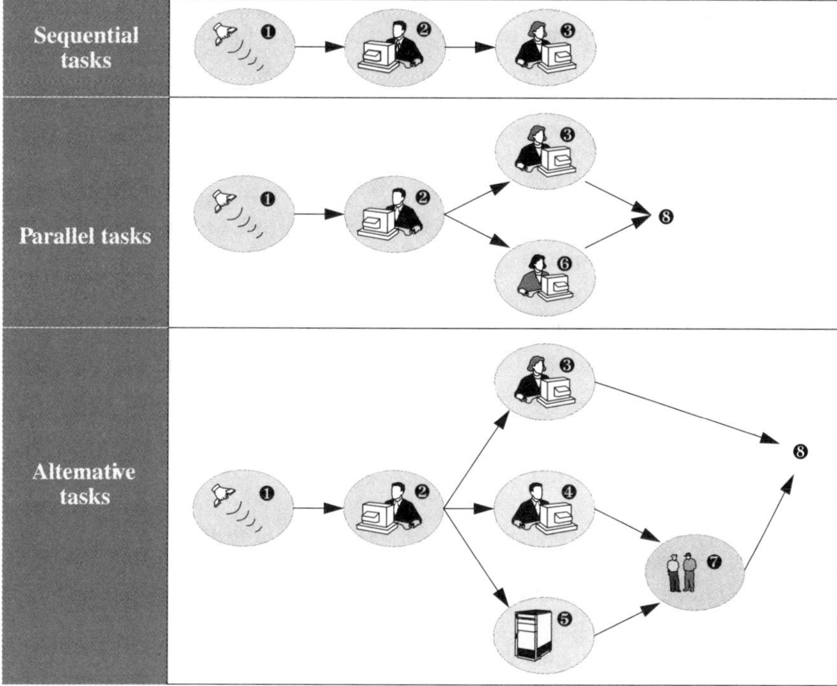

Figure 11.6 Transactional semantics in a service provisioning workflow.

The customer and service management business processes described in the Chapter 10 are well supported by the workflow technology. In the next sections, we further explain how the workflow technology can be combined with CORBA distributed object technology to implement business processes and to support the interaction logic between the processes.

11.4 CUSTOMER ACCESS COMPONENT

The customer access framework partitions the functionality of the customer access mechanism into five subsystems: the terminal, the gateway, the presentation logic adapter, the process workflow controller and the application logic adapter. The terminal is a pre-exiting UI technology that the application must accommodate. The other subsystems are specific to the customer access framework. In this section, we discuss the design and interaction of these subsystems.

A customer application interacts with the customer management system using its selected access technology. The access to the customer care management system is achieved through an appropriate presentation logic adapter (PLA). The PLA is responsible for technology conversion. For instance, some customers may prefer to report their problems by making a phone call, while others may wish to use the Web to search for the desired services.

In the customer care management system, each type of customer access is viewed as a well-defined business process workflow. This workflow consists of certain tasks with state transition rules to link them. The transition can be triggered by business events, such as customer interactions and the results of the access to the services.

A complete life cycle of a customer access is a complete transition of the business process workflow conducted by a business process workflow controller. During this transition, the controller interacts with customers through GUI such as HTML pages, which may represent an ISDN service order form in HTML page format defined by the service provider, or voice menu and prompts, which may represent the telephone bill payment options. These interactions are performed through a set of presentation logic adapters. The workflow controller also interacts with the application logic and access logic such as retrieving a customer record from the database or provisioning a teleconference connection by invoking service management layer functions. These are achieved through application logic adapters.

An implementation of customer access services in this architecture includes the following software modules:

- An instance of business process workflow controller for each customer service, this forms an application template for the specific process.
- A module to support a specific GUI technology (voice or HTML) and all the PLAs for the workflow controller to connect the access logic to GUI.
- A set of application logic adapters (ALAs) to encapsulate each application and resource access logic.
- A set of PLAs, ALAs, and associated GUI to support customer service management function.

The implementation of the customer access component provides a set of tools to define business process workflow and to generate the instances of the workflow controller. The tools also generate PLA and ALA templates to integrate with the customer GUI and application logic.

Many of the subsystems, especially the workflow subsystems, such as the business process workflow designer, business process workflow controller, etc., are not only used for developing customer care system; they can also be used to develop service management systems. In the next section, we discuss how these workflow subsystems are used for service management system development.

11.4.1 Systems Design Overview

The development process for customer access component [22] is illustrated in Figure 11.7. The development process contains two design tools for customer access applications: a workflow designer and a UI layout designer. Inputs to these design tools are the configurable rules specific to applications. The workflow designer takes the detailed business process definitions and workflow transition rules as the input and generates an instance of workflow controller specific to this process. It also generates an interaction model used by the UI designer to produce presentation logic adapters. The UI layout designer takes the UI layout rules and the interaction model generated by the workflow designer as the input and produces presentation logic adapters for the workflow controller GUI and customer GUI (HTML pages, voice prompt, e-mail, etc.). The shaded boxes represent modules generated by the development tools, and the transparent boxes represent modules specific to the application which have to be implemented individually.

The development process starts with the analysis phase, based on the understanding of the application's business processes and requirements. The design phase produces the process

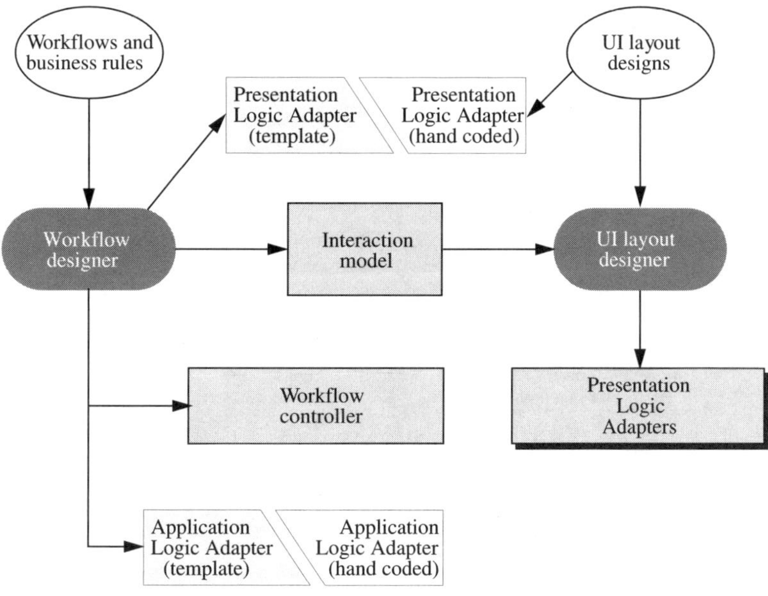

Figure 11.7 Customer access development overview.

workflows, the business rules governing customer interactions, and the user-interface layout designs.

Two development tools, the workflow designer and the UI layout designer, use the output from the analysis phase to create formalized graphical representations of these design concepts.

The workflow designer allows the developer to create the interaction model based on the process workflows and business rules. The workflow designer also generates the code for the workflow controller and the templates for the application logic adapter and presentation logic adapter. These templates can be used by application developers to produce the complete application logic adapters and presentation logic adapters.

The UI layout designer assists with the development of the presentation logic adapter. It allows the designer to configure the interaction model's presentation details for generating a complete presentation logic adapter.

11.4.2 Business Process Workflow Designer

The business process workflow design tool allows service providers to model the service's business logic and the business events to perform the state transition rules and customer GUI integrations.

Figure 11.8 depicts a design example of a specific on-line trouble report business process logic with both manual and automated tasks. When the design is completed, the tool generates an instance of the workflow controller for this service.

A developer starts the workflow designer tool and enters the business rules for the application. These rules are entered through a visual environment where boxes are created to represent processes and interconnecting lines to represent state changes. The state of the workflow is made persistent as the developer creates it. This state later becomes the input to the run-time system software.

Once the workflow has been designed and reviewed, the code can be generated to automatically produce the template application. Figure 11.9 depicts the boundary of generated entities by the designer.

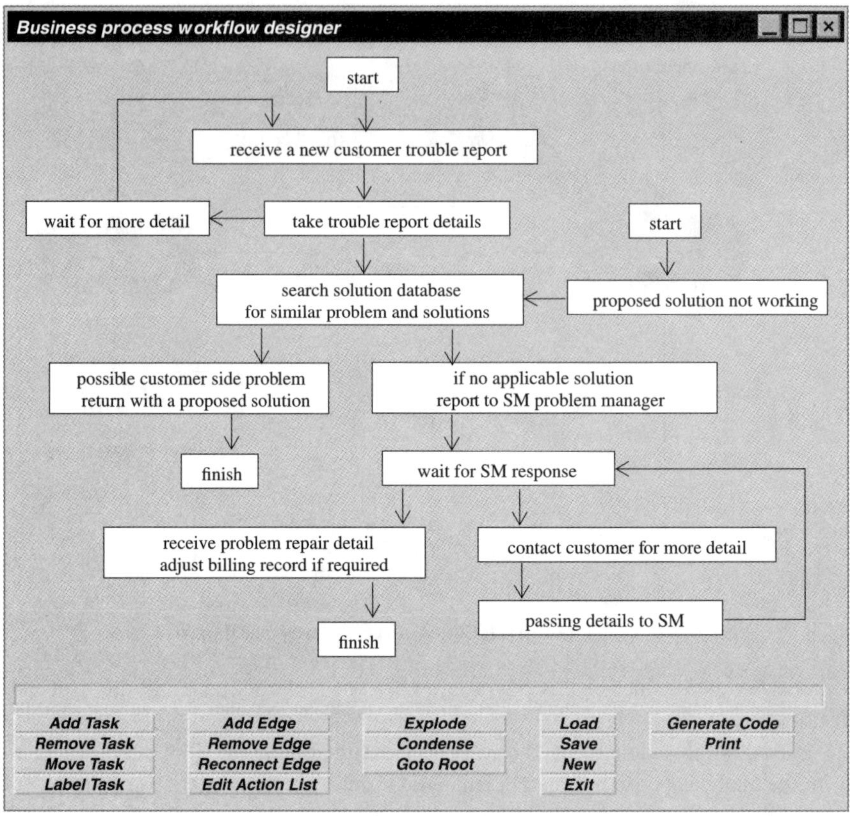

Figure 11.8 Business process workflow designer GUI.

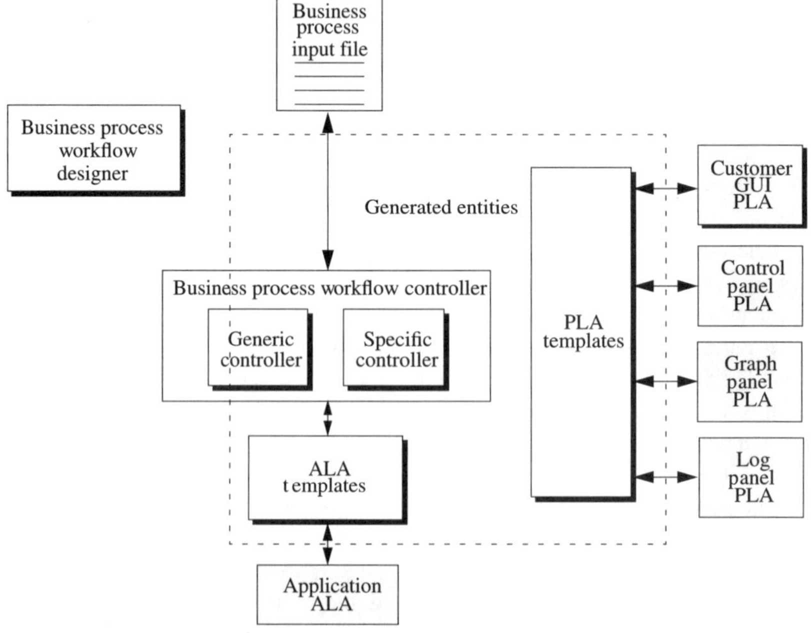

Figure 11.9 Business process workflow designer architecture.

The following entities are generated by the business process workflow designer:

- A specific workflow controller that implements the business process rules. This controller is a realization of the control semantics defined by the business rules (state transition rules, business event propagation rules, etc.). The controller also conducts the process interactions between applications and presentation through generated adapter templates.
- A set of application adapters which interacts with the underlying OSS systems or network management systems.
- A set of presentation adapters which interacts with the customer designed GUI, such as displaying an HTML page or download customer input from Java applets.
- Three specific presentation adapters that handle the presentation of workflow controller GUI. The customers can interact with a running instance of the workflow controller through these three GUI frames.

11.4.3 Business Process Workflow Controller

The generic workflow controller links PLAs to the ALAs through a controlled state machine.

The workflow controller mediates in all interactions with a customer for a given service. Each workflow controller has knowledge of the workflow state, such as which state each user session is in and what transitions from that state are possible. When completing a state transition, the controller calls an appropriate method on the ALA interface to service that state transition. During the execution of a workflow, the controller ensures that the system can only transit into a legal state and cannot become unstable.

The PLA provides the user interface to the system. For the workflow controller, the user interface consists of three panels—graph panel, control panel, and a log panel. The graph panel shows a representation of the workflow and its state through a color coding. The control panel shows all actions customers can perform at the current state. The log panel provides a short text description of the current workflow status. These three frames offer a management GUI which can be used to supervise the execution of a workflow.

The PLA also provides the integration with customized GUI. For instance, the workflow business rules may dictate that when the state transition reaches the task *processing payment,* the customer defined HTML payment page will be displayed. The workflow controller will call the display of this GUI (with a predefined handle) when this task is invoked.

The ALA is the glue between the business rules and the actual service implementation in the service management layer. Each defined interface method on the ALA may be implemented by multiple calls to multiple underlying service components. For example, the `TroubleReport()` logic adapter method may perform the following steps: searching solution database, invoking problem management functions in the service management layer, and then communicating with the customer for the results.

Figure 11.10 depicts the run-time business process workflow controller GUI. The right-hand side shows three panels for management purposes.

11.4.4 Customer Access Application Development

The customer access system can be applied to a number of areas where a customer service provided by a service provider can be represented as a business process workflow. Within a business process, the complete steps of interactions between a customer and the service provider forms an instantiation of the workflow. These customer access processes include:

- On-line service order
- Customer care system and customer quality of service management
- On-line customer access such as billing inquiry

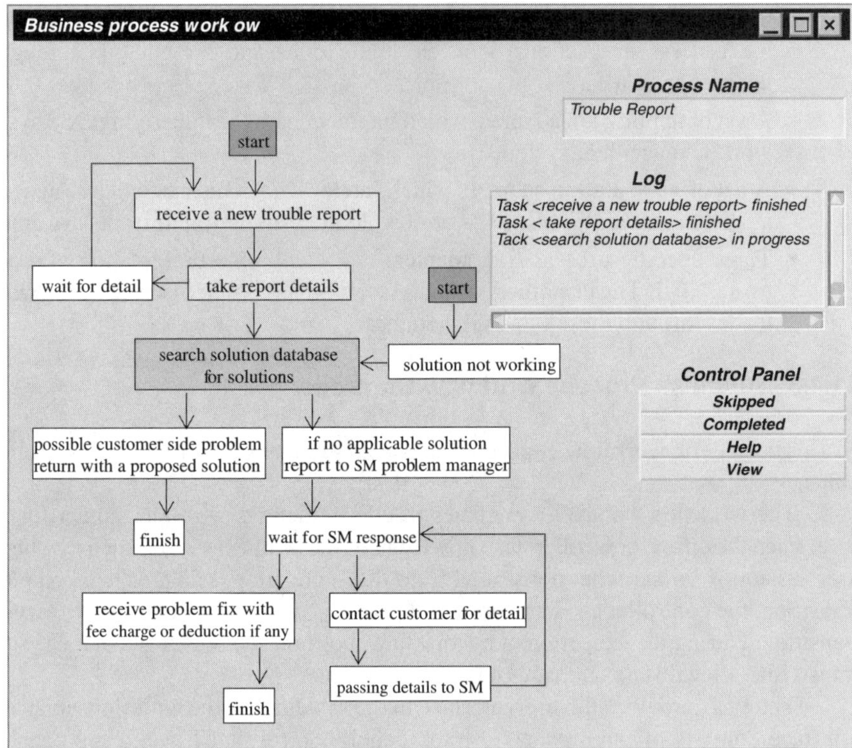

Figure 11.10 Business process workflow controller run-time GUI.

The business process workflow controller and its GUI as depicted in Figure 11.10 can also be used to provide the customer access management function. This GUI can be invoked both at customer side and at the service provider side to monitor the execution of the instances of the customer access. The controller keeps all log information in persistent storage which includes all completed state transitions. This enables customers and service providers to replay the customer interactions even after the service is provided.

The controller can also be extended to integrate with the application's quality of service management functions so that it represents a more general customer service management tool.

The software architecture presented here has been used to develop real operational customer access systems. A system based on this architecture and tools has been developed which enables the Web-based on-line customer ordering management and customer access to billing systems for call record and bills.

11.5 SERVICE MANAGEMENT COMPONENT

The software architecture for integrated service management supports service ordering, service provisioning, service assurance, and billing functions. The architecture supports a scenario where multiple customers through different customer access and customer care components interact with multiple service providers. Customers select services among different service providers, and service providers supply services for many customers. Service providers support a distributed and federated product and service catalog that provides the on-line trading and location of services. Figure 11.11 depicts this architecture.

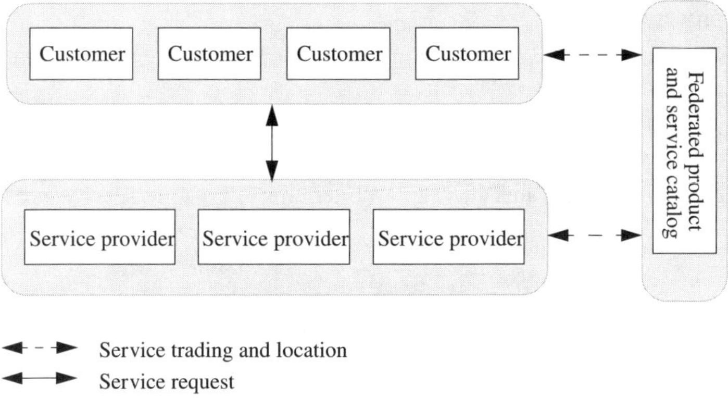

Figure 11.11 Global service environment.

Each service provider may register themselves and their services with the catalog. Customers can search the catalog for services and service providers. The customers then communicate with the selected service provider directly for service requests, or they may use their favored service provider to access services on their behalf.

The catalog can also be used by the service providers to locate and trade services between themselves in order to support the end customers in a one-stop-shopping scenario or to provide value-added services. When a service provider (e.g., service provider A in Figure 11.12) requires subservices from other service providers (e.g., service providers B and C in the same figure) it can directly communicate with them through the search of the catalog.

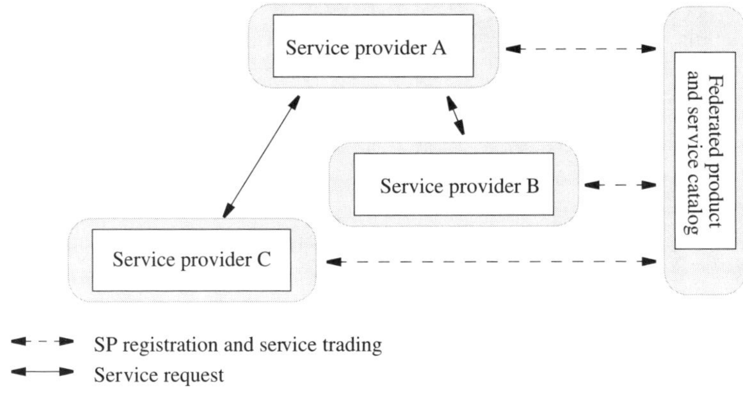

Figure 11.12 Service location and trading between service providers.

11.5.1 High-Level Architecture

In the software architecture, the following software components are identified in each provider's service management environment:

- A client software component representing the customer access capability supported by a service provider. It contains a customer access interface and customer care management system. The examples of this component include the call center software, electronic customer access, software and other components used by the customer service representatives.

- An SP server software component representing the service management capability of the service provider. It supports the service access interface and service management processes.
- A number of integration software components supporting different types of integrations.
- An in-house or outsourced network management system.

Figure 11.13 illustrates such an architecture in a single service provider environment.

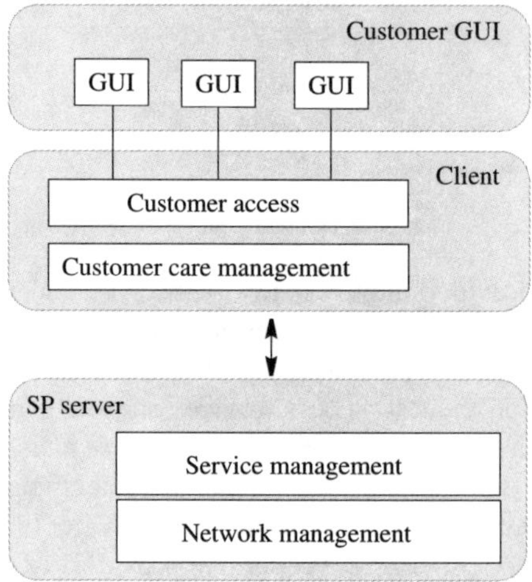

Figure 11.13 Service provider architecture.

Details of the client software have been discussed in the Section 11.4 as the customer access component of the integrated management architecture. This section focuses on the SP server software details.

11.5.2 SP Server Architecture

The main functionality of the SP server is to support the service management business processes and their integration with the business objects. It also supports the interconnections between peer service providers. Figure 11.14 depicts an implementation architecture for business processes. In this architecture, a workflow design and generation tool is used which takes workflow specifications as input and produces instances of workflow controllers as output. A workflow controller is an executable CORBA server that interacts with business objects defined as part of the service interaction model. These business objects are defined in this architecture using CORBA.

Each service provider is implemented as an SP server containing the following subsystems:

- An `SP` object that represents the service provider and is responsible for administration aspects of a service provider. It controls the portion of the service catalog the service provider is offering. The `SP` object is also the initial contact point of the service provider for all customer interactions.

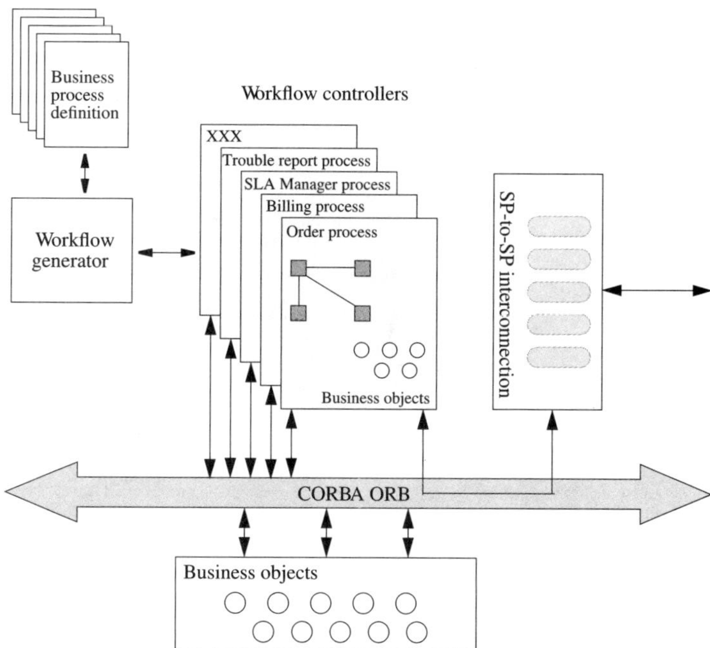

Figure 11.14 Business process architecture.

When a customer requires a service, it communicates with the selected service provider via its initial contact point supported by the `SP` object. Upon receiving a customer request, the `SP` object creates a service manager instance to handle further communications with this customer.

- A `service manager` which is an component responsible for all service management requests issued by a specific customer. One `service manager` instance is created by the `SP` for each customer access session. The life cycle of the `service manager` is the whole duration of the customer access session. The `service manager` interacts with all service management processes for customer requests.
- A number of service management subsystems and components which implement the business processes presented in Chapter 10. These components are business process workflow controllers generated by using the business process workflow designer. The integration with other service management processes and the integration with network management components are defined as part of the process. The interworking with other service providers is supported through various OSS interconnection gateways which communicate with the peer service provider systems. These peer systems may or may not use CORBA platform.
- A number of service and network management integrators which are responsible for the integrations between the service management layer and network management layer within one service provider environment.
- An SP-to-SP interconnection component that is responsible for interworking with other service providers.
- A data repository that stores all service data including service objects, SLA records, performance data, billing record, alarms and faults, etc.

Figure 11.15 depicts the interactions between different components. These interactions reflect the logic relationships between different processes. In a computational sense, all communications are invoked through the common computational architecture, i.e., the objects are invoked through ORB.

In the remainder of this section, we examine the design and implementations of some business processes in more detail.

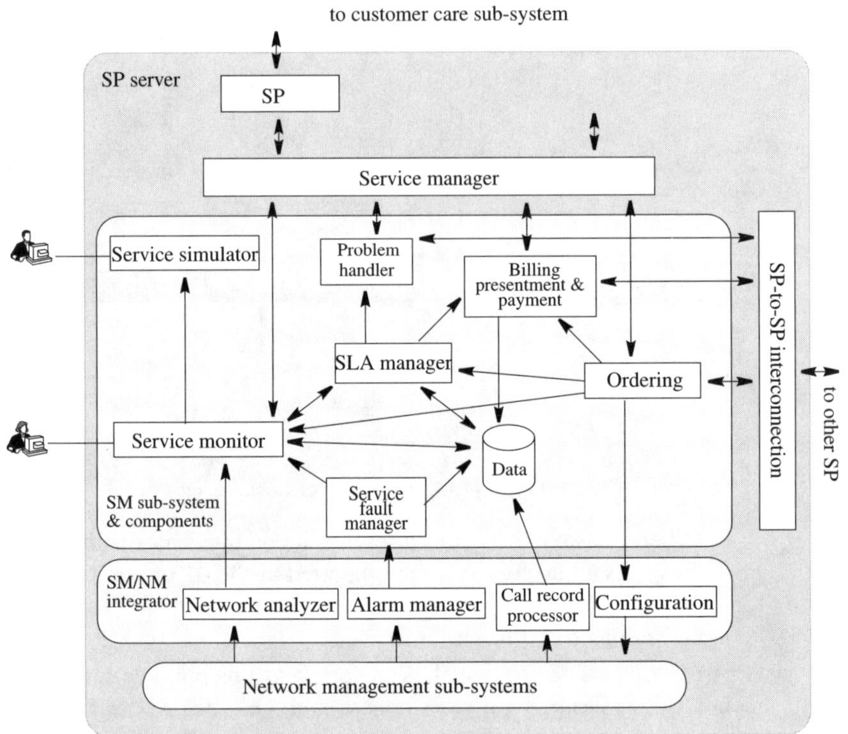

Figure 11.15 SP server integration architecture.

11.5.3 Service Ordering and Provisioning

The service ordering and provisioning is one of the major service management processes. The ordering process contains features for service parameter negotiation, provisioning, SLA management, and service delivery. It includes service provider workflow supporting process flow-through, and an interworking based on the service interaction model.

11.5.3.1 Order Scenario. In the service ordering scenario, the customer orders a service from a service provider. A sole contact point and a set of integrated management functions are provided to the customer through the `SP` object. The service provider may in turn order subservices from other service providers to deliver a packaged service to the customer and facilitate all interactions with other service providers on behalf of its customer. This extended subservice ordering is transparent to the customer. In this process, there are a number interactions with the business objects concerning the service ordering, such as a service offer, service quality parameters, an SLA, a service order, etc.

Figure 11.16 depicts the service ordering information flow at both customer side and service provider side. A customer can send a service order request to the service provider based on the service catalog search. The customer can then negotiate with the service

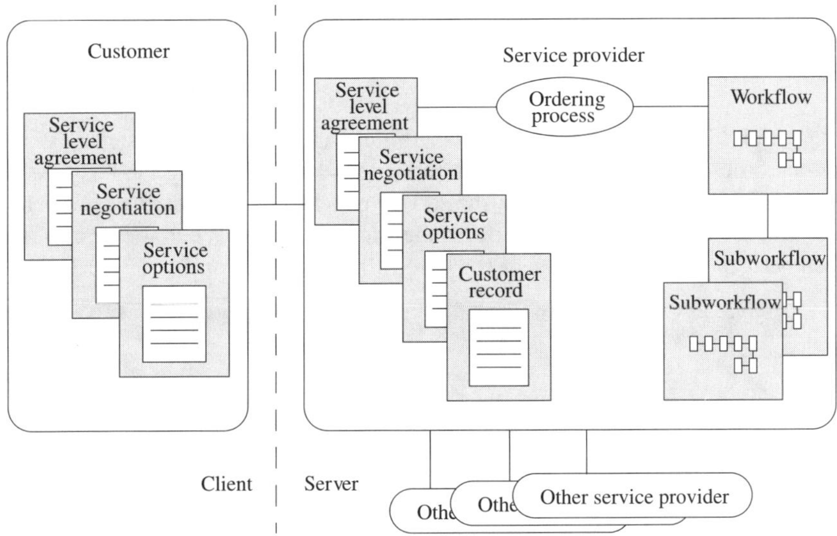

Figure 11.16 Service ordering scenario—information flow.

provider in multiple sessions to reach a satisfactory order and agreed SLA. These negotiations cover the service features and QoS parameters based on the service offers and options from service provider.

On the service provider side, service offers and options are provided to the customer for negotiation in order to achieve a final service order and an SLA. The service provider uses the service order workflow controller to control all the internal service ordering and provisioning tasks and to coordinate all interactions with the customer and the internal fulfillment activities. The service provider can use exactly the same interface to require subcomponents of the service request from other service providers as part of the overall service offer. The workflow controller follows the business process specifications for service ordering and monitors the status of ordering tasks.

Some tasks in this workflow can have their own subworkflow processes if the tasks can be further decomposed into other workflows. An example is the development of an order plan which includes subworkflow activities of interworking with other service providers for subservice ordering.

The workflow controller plays a very important role as it provides connections between visual representation of the ordering and provisioning process and the fulfillment activities and controls these interactions.

The service interaction model is used as the interface between a customer and a service provider. It is also used as the interaction model between different service providers. There are no fundamental differences between the SP-to-SP relationship and the customer-to-SP relationship, since in any SP-to-SP ordering interaction, one service provider is playing the role of customer. The differences appear in the granularity of the ordered services and negotiation process. At the interface between a customer and a service provider, the system is capable of handling complicated packages of services (e.g., including voice, data, Web) and supporting ad hoc customer requests and negotiation. However, at the SP-to-SP interface, the system mostly handles well-structured simple and predefined services (such as leased lines, VPs, etc.) and the SLA is likely to be a derivation from the overall business agreements between two interworking service providers.

In this implementation, the same service interaction model is used to support both customer-to-SP and SP-to-SP interactions. In a one-stop-shopping situation, these ordering interactions are recursive. Thus, the strategy to treat the SP-to-SP relationship as an instance of customer-to-SP relationship is very useful for increasing the software reuse and interoperability.

11.5.3.2 Service Order Process. Figure 11.17 depicts a design of a specific on-line service order logic with both manual and automated tasks using the business process workflow designer. Using the designer, the service provider can model its business logic of the service and the business tasks to perform the state transition. When the design is complete, the designer generates an instance of the workflow controller for this service.

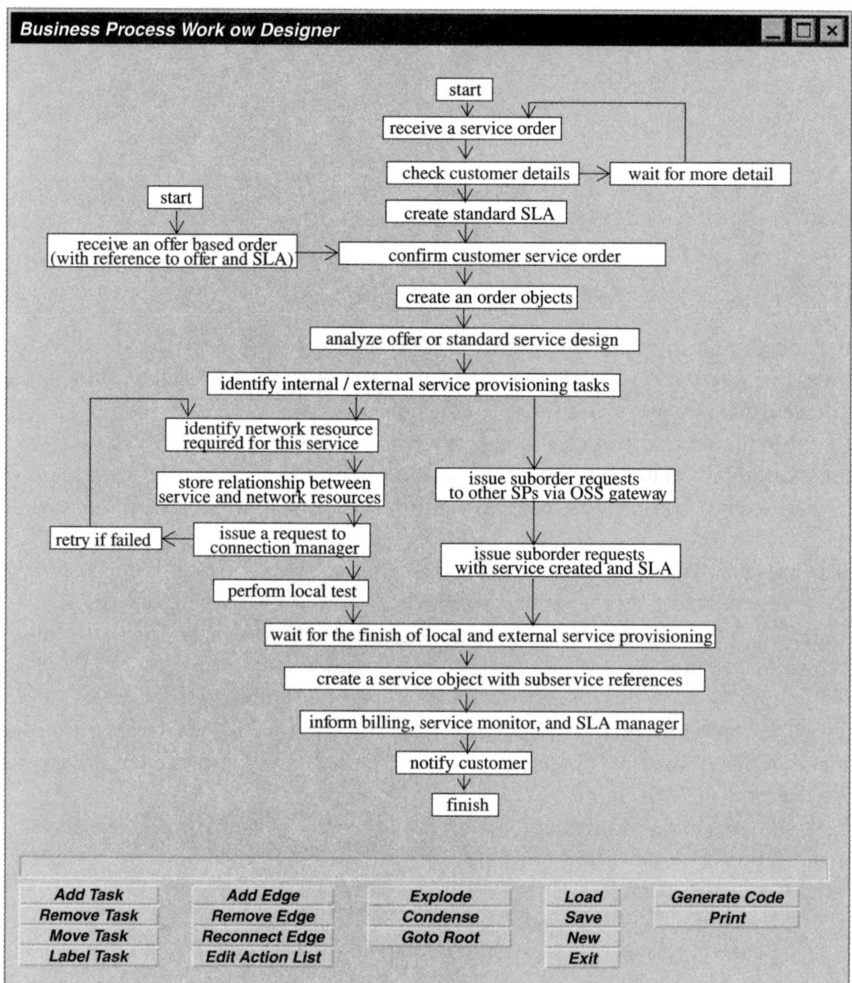

Figure 11.17 Service order business process design.

This design captures a number of business rules in providing the required service. For instance, if the service provider is a local carrier while a customer requires long distance services, the business rules define which long distance carriers will be used to support the services.

The application logic adapter (ALA) described in the previous section is the glue between the business rules and the actual service implementation. Each defined interface method on the ALA may be implemented by multiple calls to underlying service components. For example, the `DevelopOrderPlan()` logic adapter method performs the following steps:

- Identifies whether the required service can be solely provided by this service provider based on the service features and the set of agreed SLA terms and conditions.
- If the service provider is capable of providing this service and meeting the quality of service requirements specified in the SLA, then invoke `ServiceConfig()` logic adapter method for the service configuration.
- If the subservices are required from other service providers, then the implementation of the ALA identifies the components of customer service order that are to be provided by different providers, and proceeds with the service configuration:
  —invoke `ServiceConfig()` logic adapter method for the service components which are to be provided locally.
  —issue a sequence of service order requests to other service providers.
  The last two tasks are performed in parallel.

The `ServiceConfig()` logic adapter method is the integration point between service management layer and the network management layer for service provisioning. It maintains the relationship between service instances and the network resources involved in implementing the services, and it is able to invoke network management functions to configure the required network resources. There are also other integration points for other management functions, such as fault management and account management.

During the service ordering process the workflow controller interacts with ALAs to access the underlying service management and network management functions. These include interacting with the network provisioning system, the SLA management system, and the billing, fault, and performance management system. These interactions are performed by making calls onto the appropriate ALA interfaces. The ALA then makes suitable calls onto the underlying service infrastructure.

The workflow displayed to the operator reflects the process life cycle. This includes the steps carried out by the service provider to implement the service order and the interworking with other service providers. The state information of the workflow is also accessible by the operator. This information reflects the dynamic status of the workflow instance representing an interaction of this process with the customer. At any stage, the customer may query about the progress on the service order, and check the service parameters against those specified in the order.

Figure 11.18 depicts the business process workflow controller GUI.

11.5.3.3 Integration Processes. In order to provide a service ordering and provisioning capability, the ordering and provisioning processes must interact with many other service management processes and network management processes. The following two types of interactions are of particular interest.

SUBORDERING. The order process in the service interaction model is defined with the purpose of support exchanging of ordering information between two service providers. In a typical ordering situation, when a customer orders a service from a service provider, the service provider may suborder part of the service from other service providers. This chain of subordering can theoretically continue to any length.

The subordering is well supported by the OSS interconnection gateway technology. The technology supports the minimum process interactions between different service providers using different management technologies.

PROVISIONING. Service provisioning offers the fulfillment for service ordering. Service provisioning component provides a service level view for the network resources. The key functionality of service provisioning is to understand the service logic and be able to map the required service onto the functions and services supported by the network management layer.

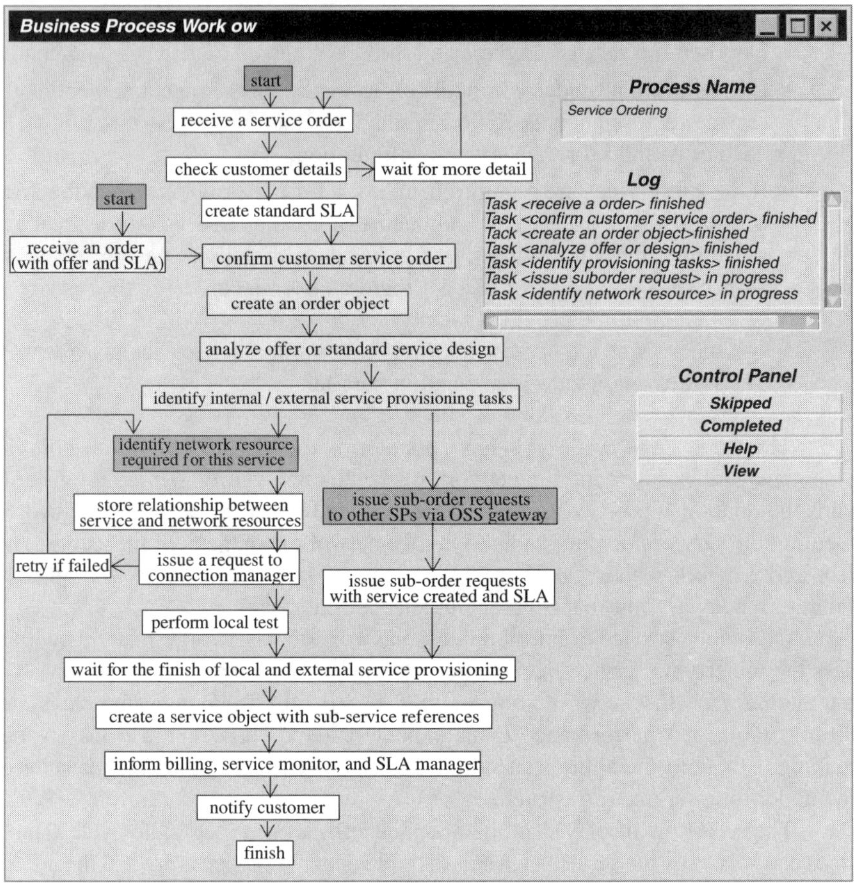

Figure 11.18 Business process workflow controller.

The service and network management integrators are software components acting as integration points between service management layer and network management layer. The `ServiceConfig()` application logic adapter used by the provisioning process is one of such integrators. The service provisioning system calls this logic adapter to access the network configuration functionality.

The network management interface used in the system supports the generic network model, which encapsulates the specific aspects of networks such as the network elements produced by different vendors, the network management protocols and stacks used, etc. The generic network model presents to the service layer a generic view of the networks and network resources which can be used to implement telecommunication services. The CORBA relationship service is used to relate the service level view and the network level view.

When viewed at the network level, a service provisioning task is often translated to the task of setting up a connection between two or more terminations. Each service differentiates itself with others by its service profile such as a set of terminations, the bandwidth required, the quality of service desired, and the existing SLA, etc. This profile dictates how the network resources are utilized to support this service.

11.5.4 Service Assurance

Service assurance components in the implementation architecture include the management of service level agreement, performance management, and service fault management.

11.5.4.1 SLA Management. The service pre-ordering process and the ordering process define an SLA between a customer and a service provider with regard to the service ordered. The ordering framework (including service pre-ordering) allows customers to negotiate the quality of service parameters in order to establish an SLA with the service provider. The service provider may suborder parts or all of the service from different service providers, thus resulting in SLAs between different service providers.

In any service provider's provisioning and management environment, the SLA management is a very important component. For customers, SLA is a way of measuring the quality of services delivered by a service provider. For service providers, SLA is the target to achieve in order to win customer satisfaction and loyalty. Thus, the SLA management is the focal point of the customer service management activities. Figure 11.19 depicts the service life cycle from the SLA point of view and the different subsystems involved in the overall SLA management.

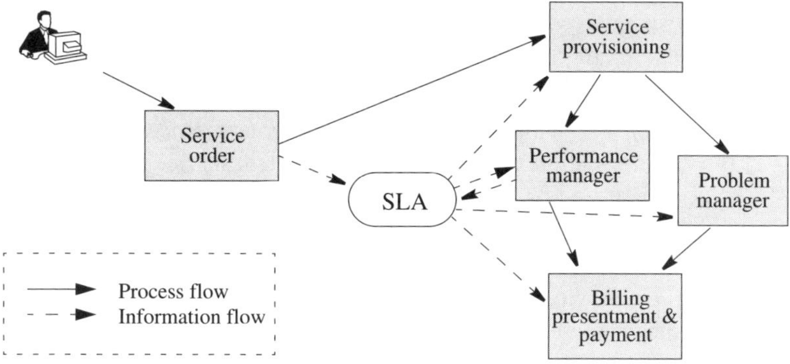

Figure 11.19 Service management life-cycle example.

The service assurance architecture supports the interactions between SLA manager and other management processes, in particular the performance manager.

11.5.4.2 Performance Manager. A performance manager contains the following components:

- *A network analyzer*—takes network performance data and network topology information to form service performance data. It passes the service performance data to the service monitor for service performance monitoring.
- *A service monitor*—monitors all the service performance parameters.
- *A service simulator*—simulates the services affected by network topology change.

NETWORK ANALYZER. A network analyzer is incorporated into the service monitor. The network analyzer takes the network performance data, such as the measurement, network usage, capacity, and congestion, and combines them with the knowledge of the network topology to form a service performance data that can be further analyzed. The service performance data are passed to the service monitor.

This component is one of the service and network management integrators. It understands the object models in both the service management environment and the network management environment and it is able to correlate service instances and network resources used to implement these service instances.

SERVICE MONITOR. A service monitor enables operators to monitor the service performance data, to compare the performance data with the quality of service parameters negotiated during the service ordering phase, and to alert any breach and potential breach of these agreed service parameters. To achieve these functions, the service monitor associates the service connection information with the quality of service information in the SLA to establish the cause or potential cause of SLA violations.

For instance, whenever there is network congestion or failure, the fault management system will capture this at the network management level and pass this information to the service fault manager. The service fault manager identifies the affected services and forwards the service fault information to the service monitor. At the same time, the network analyzer may also pass the network congestion information or network element breakdown information to the service monitor. The service monitor reflects all collected information and identifies any potential degradation or breakdown of services affected by this network fault.

The service monitor performs the following activities:

- Receives service connection information, such as creation, deletion and update, from the service ordering process and stores this information
- Generates service usage data and failure rate, etc.
- Communicates with other service management processes for other service performance information such as problems reported, service affected by network fault, etc.
- Checks collected performance data, and if required, retrieves SLA information to identify an SLA violation. If a violation is identified, it reports to the SLA manager.

It is also the service monitor's responsibility to produce the performance report to the customer via the customer care subsystem.

These processes are involved in many stages of a service life cycle including the service ordering, service provisioning, service activation, and service assurance. The following service management processes are required to provide the data necessary for the service monitoring:

- SLA manager
- Problem manager
- Service ordering
- Service fault manager

Figure 11.20 illustrates the relationship between these processes. An automated service reconfiguration system can be developed based on the service performance data generated.

SERVICE SIMULATOR. A service simulator performs the network and service simulation and scenario management function. The service operators can use this feature to simulate changes in network traffic and administrative/operational status of any connections and network equipment to determine the impact of this change to services carried by the network. For instance, the operator may *disable* a switch in a simulated network topology display to get a report of any services relying on this switch. This allows the operations and maintenance personnel to conduct an analysis for a *what if I take this equipment out of service* scenario. This feature can also be used as a network capacity planning tool.

The understanding of the service integration model by these modules and the use of CORBA as the integration platform enables these modules to be built and deployed to interact with each other. In the deployed application systems, they interact with each other through the CORBA platform and each has a shared view of the service interaction model enabled by the CORBA relationship services.

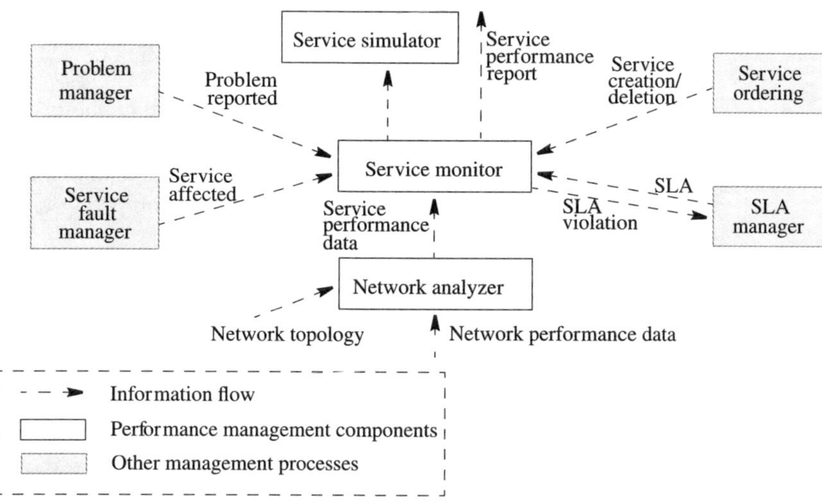

Figure 11.20 Service monitor and other processes.

11.6 CORBA AND BUSINESS PROCESS WORKFLOW INTEGRATION

Our experiences showed that CORBA, as the distributed object computing technology, provides a very good integration environment for business objects. The experiences also showed that workflow technology provides a very useful environment for business processes integration. The integration of these two technologies become crucial to integrated application development.

CORBA provides a distributed object infrastructure for service management systems and supports interactions between service management system components. This infrastructure can also be used by transactional workflow management systems to coordinate and control service management processes.

CORBA itself does not provide workflow tools (workflow engine, designer, monitor, etc.). It requires a workflow system either as an application service or as part of the infrastructure so that a distributed application can use this service or infrastructure. CORBA's underlying ORB communication mechanism, its object services, in particular the object transaction service, provide an environment to integrate with the workflow technology.

Our experience shows the following integration and application development issues with using both object transaction service and the workflow tools.

NESTED TRANSACTIONS. Most existing transaction processing technologies are based on a flat transaction model with ACID property, which does not support nested transactions. In the CORBA transaction model, where nested transaction management is supported, there is very little support at the resource management level for nested transaction semantics.

Nested transactions support commitment of a subtransaction within a transaction and allow the subtransaction to be rolled back later. These transactions are useful for the following purposes:

- *failure containment*—In the flat ACID transaction model, failure in any of a global transaction leads to the rollback of the whole transaction. This is very inflexible behavior, and very inefficient for workflow applications to implement service management processes. In many cases, failure in a task should be contained locally, that is, within the smallest subtransaction where the failure occurs. This prevents the whole workflow being rolled back. The workflow manager should be allowed to re-initiate the same task at a different time (or with different parameters) or to initiate an alternative task.

- *alternative subtransactions*—This feature relates to failure containment. It is useful when the workflow manager wishes to start an alternative task (an alternative ordering from a different source) after a task performing a similar function has failed; or when the workflow manager wishes to initiate a set of competing tasks in order to achieve a commitment according to certain semantic criteria.
- *parallelism*—This feature allows the workflow manager to explore maximum parallelism by initiating multiple tasks in parallel as subtransactions.

ROLLBACK BEHAVIOR DEFINITION. The existing transaction model implemented in most commercial software products is based on ACID property; that is, a transaction is defined as a set of tasks forming an atomic action. This *atomicity* of a transaction must be maintained at all times. The property of atomicity means that any data operated upon by the actions within a transaction can only have two states: an *initial* state at the commencement of the transaction, before any operation is performed, and a *final state,* when all required actions are successfully performed. No other data state is allowed access by the outside world of a transaction.

The property of atomicity is very desirable and relatively easy to achieve in database-oriented applications. The atomicity of a transaction is the guarantee of data integrity. However, in many applications involving non-database activities, such as human tasks, and control of processes and equipment, it is very difficult to follow this model. Examples that do not fit the model include:

- A provisioning task may involve workers digging a trench in order to lay a cable. When a rollback is required, it is very undesirable to restore to the initial state.
- An operation resulting in certain behavior of a piece of equipment (e.g., printing a check) may be very hard to rollback to the initial state.

In both cases, it is desirable to allow users to define rollback behavior of objects and operations. The reasonable behavior for rolling back the task of laying a cable may entail the work completing, and an update to the asset database to record that the customer premises cabling equipment exists, although the provisioning workflow may have been rolled back. In the second example, the operation to destroy the check may be regarded as an acceptable rollback definition.

COMPENSATION ACTIONS. Compared with rollback behavior which deals with how to rollback a task, *compensation* deals with committed tasks. There are many tasks that do not support nested transaction semantics; that is, after they are committed, it is impossible to roll them back when the whole workflow needs to be rolled back. One option in this case is to start another task (subtransaction) whose actions compensate for the previous committed task. An example of compensating for an entry in a telephone directory when the provisioning process failed is to remove the record from the directory.

Compensation can be used as an alternative mechanism for supporting rollback of committed subtransactions within a global transaction. The ACID property cannot always be guaranteed in this case. Similar to rollback behavior definition, the compensation needs to be specified as part of the task specification of the workflow.

SPECIFICATION OF SEMANTIC DEPENDENCIES. There is no provision in the existing transaction model to define temporal ordering between different tasks within a transaction. Also, there is no mechanism to specify semantic dependencies between groups of operations in a transaction. Both features are important to workflow applications.

A workflow normally has a strong temporal logic associated with its tasks. Certain tasks cannot commence before some tasks have been committed or rolled back. An important part of the design of a workflow is to specify this temporal logic.

There are semantic dependencies between workflow tasks. Some tasks must be performed in sequential order; others are performed as alternative but competing tasks—that is, only one successful result is used; and some tasks are perform in parallel to achieve maximum performance.

All these features need to be integrated with the transaction semantics of a workflow, and they can be best supported with the concept of subtransaction and a definition of rollback behavior. The specification of these constraints form a major part of workflow design and the enforcement of these constraints form a major part of the workflow controller.

Task Specification. A workflow is a sequence of tasks. Each task is defined as an object or a cluster of objects. The workflow itself can also be defined as a high-level business object. The CORBA relationship service can be used to specify the containment relationship between a workflow process and its tasks and dependencies between tasks. This information can later be used by the workflow controller to determine the constraints for task invocations.

The relationship service is also useful for maintaining the relationship between a task and its compensation action. This can later be used by the workflow controller and the transaction manager when compensation is required for a previously committed task.

11.7 NETWORK MANAGEMENT ARCHITECTURE

The generic network model is used as the interface model for the network management layer. It is also used as the integration model for different network management technologies used to implement underlying specific network management system. The computational objects in the generic network model are defined as CORBA objects. The communication to these objects is achieved through CORBA object method invocation.

The underlying network management systems use CMIS/CMIP as the management and communication protocol. The key technical issue in implementing this network management layer is to provide the CORBA/OSI integration between the CORBA implementation of generic network model and the OSI network management systems.

A number of CORBA/OSI integration approaches have been discussed before. In this software architecture, the abstract object mapping approach is used for the integration between CORBA-based generic network model and OSI-based managed network objects as this strategy is more appropriate than other methods in this implementation architecture.

11.7.1 Architecture and Components

The network management architecture consists of the following major components and features as depicted in Figure 11.21.

- *OSI agent environment*—it contains a number of OSI agents and managed networks and network elements. Access to the agent functionality is through OSI manager using the OSI communication stack. The agents also use the OSI platform (stack) to report events through event forwarding discriminators (EFDs) created as part of the SMI and SMF.
- *OSI manager environment*—it contains a number of OSI managers developed as part of generic network model integrators.
- *Generic network model integrators*—these integrators form the gateway between the CORBA management environment and the OSI management environment. Integrators are implemented as CORBA objects, and they act as OSI managers in net-

work management activities. The integrators use OSI CMIS/SMF to access OSI agents via the OSI communication platform and use CORBA to communicate with other service management components.

- *CORBA factory*—it is a standard CORBA object which supports dynamic creation of CORBA proxy objects for direct GDMO access.
- *Manager/agent distribution is performed by OSI.*
- Objects in generic network models are implemented as CORBA objects. They call specific integrators to access OSI network and element management functions.
- Generic network model objects and service management functions are distributed by CORBA.

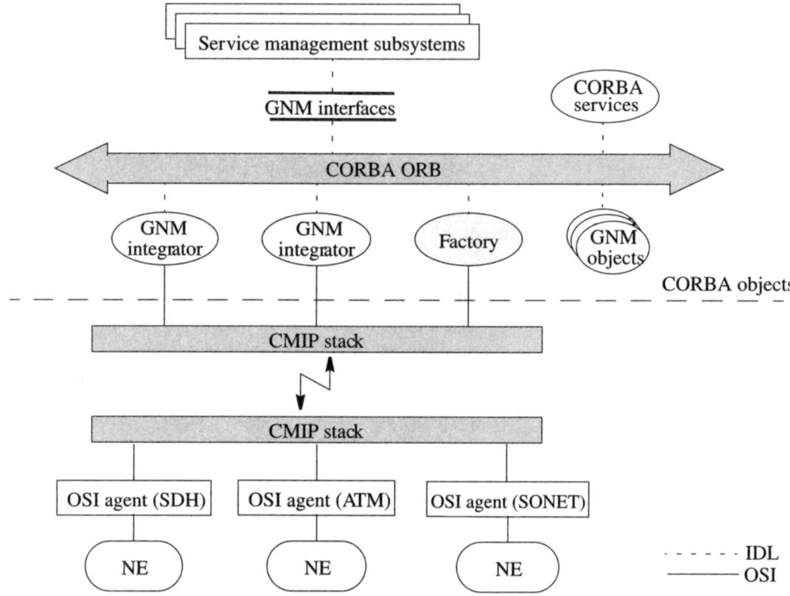

Figure 11.21 Network management architecture.

This architecture is characterized as follows:

- Generic network models are used as the interface objects to the network management layer.
- The explicit OSI management layer is supported. This function can be performed by using a commercial OSI product (OSI stack or platform supporting TMF's TMN C++ interface as an example).
- Implementation of the integrators invokes OSI manager functions. This includes:
  —mapping generic network model operations to a set of OSI operations according to the abstraction mapping rules defined during generic network object modeling stage,
  —implementing CMIS operation semantics of a GDMO object (e.g., create a connection),
  —using OSI stack's manager–agent communication style to issue CMIS requests (operations). The integrators access OSI stack through OSI management API.
- The integrators act as the gateway between CORBA and OSI environments. This gateway does not perform one-to-one mapping between generic network model objects and GDMO objects. Instead, these integrators map each abstract network

operation into a set of CMIS operations on GDMO objects (as in the example of specializing a generic network model to an OSI network model in Chapter 10). These integrators are defined as OSI management procedures, and their communication to the agents are distributed by OSI platform. These integrators have CORBA interfaces that allow them to be invoked by generic network model objects. This adds to the flexibility for migration purposes.

- The OSI platform is used to distribute OSI agents. This improves the interoperability with other OSI-based management systems.

11.7.2 Communication Model

Figure 11.22 depicts the communication model of the architecture. This architecture is based on the use of OSI manager–agent communication framework to deliver the management functionality for the element layer and network layer.

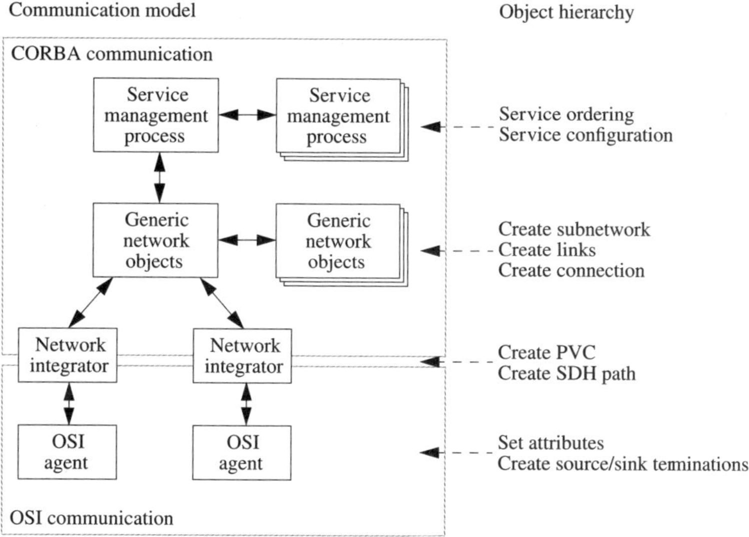

Figure 11.22 Communication architecture.

The integrators are developed as part of OSI management entities with the north-facing IDL interface, and these entities can be distributed by using CORBA. Service management layer functionality is integrated with network management layer functionality through the use of these integrators. Examples of such integrators are the configuration manager and the network analyzer described in the previous section.

One benefit of this communication infrastructure is the integration of different network management domains. At the generic network model level, both ATM and SDH can be treated as subnetworks of logical networks if required. Services with an end-to-end view can be provided across these subnetworks.

This architecture can be extended using different network management technologies, such as SNMP. More integrators are required for different network management technologies.

11.7.3 Distribution and Deployment Model

The network management architecture supports a high level of distribution. This distribution and deployment model is depicted in the Figure 11.23, which shows two levels of distribution using both CORBA and OSI. There are three entities that can be distributed:

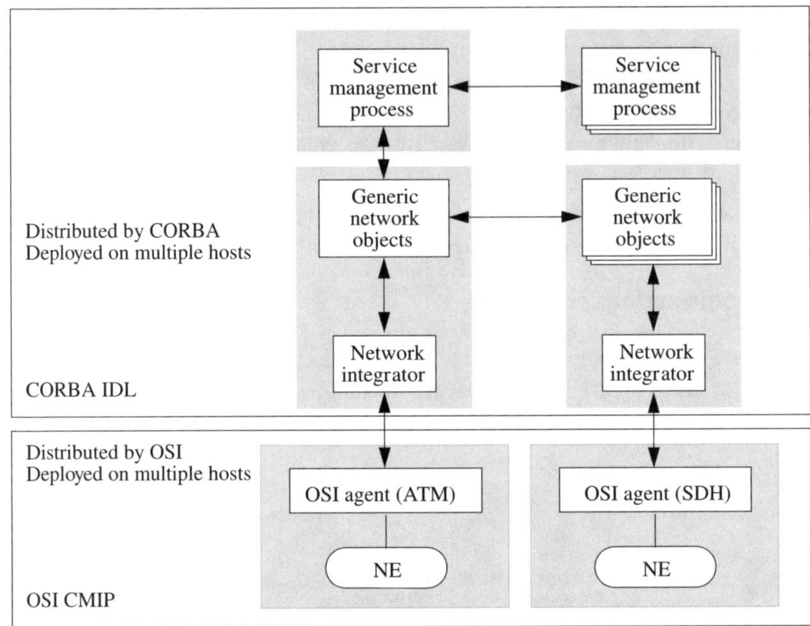

Figure 11.23 Distribution and deployment model.

- Service management processes and cross-domain network management processes (generic network model) developed and distributed by using CORBA.
- The integrators which support CORBA IDL definition for generic network model but interact with the underlying OSI systems.
- The OSI management systems (managers and agents for subnetworks).

The distribution model allows the encapsulation of the whole OSI environment (itself may be distributed by an OSI platform) by using integrators as CORBA wrappers. The same principles can be used to encapsulate other technologies such as SNMP.

For distributed deployment, CORBA objects (representing service processes and generic network model) could be deployed separately on different hosts depending on the application needs. Typically, a CORBA object representing the management functions of a subnetwork should be deployed on a separate host. Therefore, different CORBA objects representing a nested subnetwork structure could be deployed in a distributed manner. This will increase the overall system throughput and scalability.

The integrators should normally be deployed as part of the CORBA generic network objects. For instance, a CORBA object representing a subnetwork which is specialized to an ATM cloud should be combined with its ATM integrator and deployed on the same host.

The deployment of the objects in a distributed environment also influences many aspects of the system, in particular, the system performance.

Each CORBA remote operation must go through the following steps:

- marshaling the object invocation
- transmission of parameters
- demarshaling (including rearrangement of byte order)
- execution of operation
- marshaling the result

- transmitting and
- demarshaling the result.

There is an extra overhead for each object request and object result. The need to invoke a large number of remote operations to perform a management function will slow down the system. A better approach is to create higher level functions with more detailed parameters and return data structures to reduce the overhead of invoking the remote operations.

Deploying integration objects on the same machine as the generic network model objects can help reduce this problem as the CORBA operation is invoked without the need for the transmission of parameters and results. Integration object and generic network object level caching can also help improve performance.

11.7.4 Reliable Distribution Model

A service management and network management system is a distributed system. An important aspect to consider when building distributed systems is to ensure that the distribution doesn't reduce the overall reliability of the system. It is very easy to build dependencies on individual systems that result in the entire system being dependent on that system. If that system crashes, then the entire distributed system fails. This clearly leads to a reduction in the overall reliability of the system.

If carefully constructed, distributed systems can lead to a substantial increase in the overall reliability of a system. The following principles help in building distributed systems that increase overall availability:

- Design service interfaces to use stateless communication where possible
- Use replication of state to provide backup services
- Plan a mechanism for change-over after failure
- Avoid single point of failure

Figure 11.24 shows the general architecture for a backup service. Clients use the trader to locate the service. In the case of failure, the clients use the trader to locate an alternative backup service.

In most cases it is necessary to replicate the state to the backup service. If the state is stored in a database, then standard database replication mechanisms can be used to replicate the state to a secondary host.

Depending on the requirements of the system, a number of variations to this basic scheme can be used:

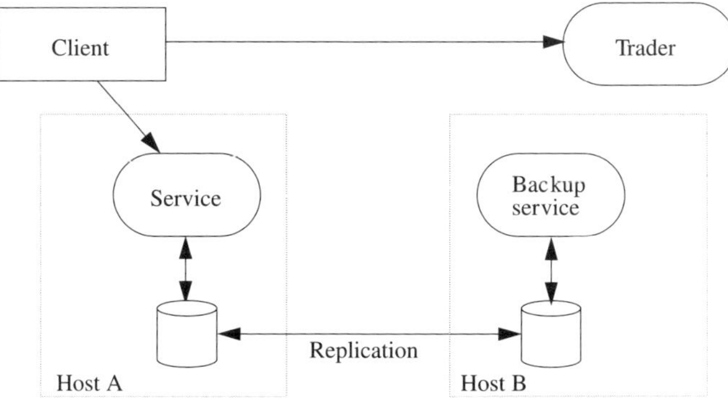

Figure 11.24 Backup service strategy.

- Use of secondary deployment during normal operations to improve performance. When a failure occurs the system continues to operate, but with reduced performance.
- Use of multiple deployment to increase availability.

A trader is typically a part of any automatic switch-over or failure recovery scheme. It is therefore necessary to ensure that the trader itself doesn't become the source of the problem. The solution to this problem is to use trader federation. Trader federation allows multiple traders to be combined to form a single trading space. In a federated trader environment, a trader will contain links to other traders in different locations. An application can direct the search to one or more traders by following these links. Trader administrators or applications can change these links as part of configuration control or update them dynamically.

When traders are federated, the failure of one of the traders only results in the loss of the offers stored in that trader. If traders are located on the same host as the objects that they have offers for, then failure of a host will only result in the loss of the offers for the object located on that host.

The replication of the service state can be either master–slave or peer replication. Master–slave replication is used where one service instance is the primary and the other is the backup. Peer replication is used where multiple service instances concurrently use the replicated state.

There are three options for replicating state to backup services:

- Database replicators
- Distribute calls to multiple servers (slows performance)
- Background replication interface

The choice of a replication mechanism depends on the particular service being replicated.

The service failures can be detected by the client application, or by the heartbeats of the services. The choice between these two options depends on the application. If recovery from failure can be achieved quickly, then it is acceptable for the client to discover the fault and trigger the change-over. If the recovery takes a reasonable amount of time, then the use of heartbeats to proactively detect the failure and trigger the recovery before a client needs it is necessary. However, while heartbeats can detect and rectify faults before clients need the service, they cannot guarantee that a client will not be the first to discover the failure.

Reliable service increases the availability of the system. If the system is designed to allow use of the backup service during normal operation (i.e., peer replication) then the performance, in terms of throughput and response time, can also be improved.

11.8 SUMMARY

The implementation strategy and the different components of the architecture presented here have been applied to produce different applications. Part of the customer service management and access architecture has been developed for a commercial electronic customer access application and deployed by some service providers to allow registered customers to on-line access service provider's billing systems. A broad range of commercial on-line customer care and access applications is being developed and deployed to support a more diversified range of electronic customer access and customer care business processes.

Part of the service provisioning architecture including the service ordering and service delivery has been deployed in a commercial on-line customer order support and management environment. The workflow for business process control was used to speed up the application development and for flexible deployment of application semantics. The integration of the service ordering, provisioning, and service assurance based on this architecture has also been developed as a large *proof-of-concept* system, and its commercialization is currently underway. This system also integrates with the underlying broadband network using the integration strategy and the generic network model. The system was developed using Java, CORBA, and OSI.

The CORBA-based implementation of part of the generic network model is currently available in a commercial broadband network management product. The product supports recursive subnetwork architecture, provides cross-network technology domain management functions, and integrates with the underlying OSI network management system. The product implemented the network management architecture described here, and it aimed to provide the cross-domain integrated network management support. The product is now been considered by many large network and service development projects undertaken by large telcos.

Our experience shows the importance of a number of key technologies in building integrated management solutions. The maturity of these technologies vary considerably, their integration toward software solutions has delivered value to the tasks of building and deploying management solutions satisfying the requirements of business process reengineering.

12

Conclusions

Building integrated management solutions to support the business process reengineering is a long-term evolving task that presents many challenges and difficulties. We have shared with the audience our experiences gained from our own commercial software solutions development and deployment, as well as R&D efforts. We hope that the lessons we learned and experiences we gain from these exercises can enlighten others who are going through the same process to build integrated business solutions.

So far in this book, we have discussed the technical challenges to telecommunication management solutions development. We have analyzed the roles of the technology and software development in the era of telco business process reengineering. We have dealt with a number of new challenges to the software solutions providers. Throughout the book, we focused on the integration of the multiple technologies and operation systems to create integrated solutions.

As we stated before, this book is not about building telco operation systems. It is about establishing development principles and processes that will lead to an integrated environment for new and existing operation systems. It is more about the paradigm and the methodology used to approach the large problem, to decompose the tasks, and to manage the development and evolution processes.

Throughout the book we used CORBA as an example of distributed object-oriented technology, and OSI as an example of network and element management technology. Although CORBA is becoming the strong technology choice for the telco industry and OSI is almost an incumbent technology for network management, we hope that the principles and processes discussed in this book will not presume that these are the only technology choices. The rapid emergence of IT technologies and their adoption in telco industry have provided a large number of choices to accomplish the same task. This has also made it impossible to predict what technology is going to win the market tomorrow. It is therefore more important to have an integrated environment that allows us to develop management applications for business purposes.

Before we close this book, it is useful to reflect on the role of software development in telco business, the things we might have learned from this book, and to look at the future challenges facing the software development for the telecommunication industry.

12.1 THE ROLE OF SOFTWARE DEVELOPMENT IN TELCO BUSINESS

The fundamental understanding of the requirements for telco business process reengineering is always crucial to the success of the solution development. This understanding should always be focused on why but not how. In today's business environment with strong competition from all different directions, telco's software solution developers cannot afford to be only technologists. They have to understand the impact of the competition, the reasons for business process reengineering and the business objectives of the reengineering.

The advancement in IT industry has fundamentally changed the software engineering practice for telecommunication management solutions development. The abundant supply of technology frameworks and tools has created business opportunities for telcos to provide the crucial value-add to their service offers and management solutions. On the other hand, this has also introduced the unprecedented complexity in develop management solutions. The software solutions development became such a double-edged sword, that so much can be achieved to help wining the competition, but so much can go wrong if it is not done properly. The software solution development and management practice for telco industry became such a special skill that telcos all over the world think now whether they should continue to be the solution developers or even the systems integrators themselves.

This has fundamentally changed the traditional value chain. Telcos can no longer rely on their own R&D departments to provide software solutions. They have to concentrate on their main business in providing better services and customer service management in order to win or just stay in the competition. In turn, the equipment suppliers can no longer just provide network equipment; they will have to act like solution integrators to provide integrated business software solutions. The role of independent software tools vendors and solutions vendors will become more prominent but they have to act like, or work closely with, systems integrators to supply solutions rather than just selling products or technologies.

This change of value chain has impacted the software solution development from technology-oriented focus to the solution-oriented focus. Since the software solutions focused on business problems are adding so much value to telcos, this has provided the software solution vendors with more opportunities. At the same time, this has challenged the software vendors to think hard about what telco industry really wants and how their software development experience can help them achieve these goals. Rather than waiting for RFPs to appear from telcos, the software vendors have to be ahead of the game, as their customers want to stay ahead their competitors.

The principles, processes, and experiences discussed in this book help software practitioners focus their thoughts on the range of business issues before they build software for telco industry. They must clearly understand the business purposes of the solutions and have the shared vision of the business objectives and strategies. They must have a clear understanding of the role of technologies, products, and standards in the solution development. Their focus must be firmly on producing long-term business software solutions rather than individual operational systems.

12.2 WHAT HAVE WE LEARNED

In this book, we dealt with the technical challenges of building integrated TMN solutions in a multiple technology environment. Although the book is not focused on building specific operation systems, it provides the principles and processes in dealing with this complex issue. We have shared with the audience our experiences in putting these principles in practice and the success and difficulties we have encountered.

It has been stressed throughout the book that the integrated management solutions have strong and urgent business requirements and these requirements, not technologies, dictate the objectives of the software development. These requirements provide reasons for the business process reengineering and strong input into the process of defining integrated solution architecture. Without the clear understanding of these requirements, we have lost the relevance and purpose of the software development effort.

The mainstream IT technologies and software engineering practices have been instrumental in supporting business process reengineering in many other industries. They have also been successfully introduced into telco industry and are playing pivotal roles in the telco BPR. These technologies and in particular the distributed object-oriented technology have greatly enhanced the strength of the TMN technology framework to face today's business challenges. However, we have demonstrated that this can also lead to a significant degree of complexity in the technical solutions and the management of TMN development. We have pointed out that before we adopt the distributed technologies, we must have a thorough understanding of the business requirements for adopting these technologies, and also a thorough understanding of the complexities and difficulties of adopting these technologies.

Many of us are not unfamiliar with the migration from centralized development environment to the distributed environment. But some failed exercises in industry suggest a lack of a fundamental paradigm shift and a lack of experiences in a wide range of new design principles and models. We always pay dearly for this lack of understanding. The industry is full of the examples of the solution development efforts turning into expensive R&D exercises. The book shared our experience to achieve this paradigm shift and to understand the right strategy and process for this migration. We have shared our view and approaches to stay focused to the business requirements and to have a clear strategy of dealing with the emergence of new technologies, new standards and new products.

Although we have shown the industry's acceptance to the notion of right technology for the right problem domain, it is not easy to put this notion to practice. We have demonstrated a process in which multiple technologies were used to provide integrated solutions, and this integration is really focused on the value-add of each participating technologies, rather than just to prove the completeness of the technology. This integration process allowed us to take full advantage of the technology wars but to stay away from their sometimes disruptive impact.

We have illustrated this process from different viewpoints. First, we focused on an approach to define the business processes and information models. We examined in-depth the commercial value of the international standards movement and their results. We stressed the need to move forward from a network interconnection-based view to an application integration and interworking view of the object modeling. Further, we demonstrated multiple types of integration strategies for different integration requirements. From this, we defined an integration framework which encompasses a well-defined business process model and information model. This framework enabled us to understand and to articulate our business better before we build management solutions. This led us to an integrated software architecture that accommodates multiple technologies and supports a better environment for solution enhancement and migration.

12.3 THE FUTURE CHALLENGES TO TELCO SOFTWARE INDUSTRY

The future challenges to Telco business software industry lie in the area of customer services and management. If yesterday's battlefield was in the area of providing better network management environment and OA&M, today's battlefield is clearly in the area of offering value-added services. The software development practice today is focused on the ability to define, deploy, and deliver new services to customers quickly. Competition from the new entrants (CLECs, mobile operators, Internet service providers, etc.) who have demonstrated the great flexibility, mobility, and adaptability in providing new services to customers will drive the solution development to focus on the issues such as:

- End-to-end service creation, provisioning, and management
- The integrated service creation and management environment
- More efficient and smooth operation of the network and OSS environments

However, what are tomorrows business challenges? How should the solution developers be prepared for these new business challenges?

Learning from the other service industry's experience, the technology and business innovation in the area of supply better services and products to customers will slow down with more and more services and products become commodity and ubiquitous. The battlefield will shift to the sales process of these commodity services to customers and the fight to win new customers and to win customer loyalty. This is already happening, but the likes of *travel agencies* and *financial brokers* of telecommunication industry have not progressed to a prominent position. Indeed, the *amazon.com* of the telecommunication services has not yet emerged, but there is no reason to believe that this is not coming. This indicates that the competition is just beginning. The programs launched by many telcos and service providers in the area of customer intimacy never existed some years ago, or if they did, they were only offering lip services. But they are all getting real now.

The result of business process reengineering, introduction of new IT technologies and tools and improved software development practice have speeded up the technology and business innovations. They allowed the telcos and service providers to take the advantage of providing value-added services to customer quickly (quicker than their competitors). At the same time, these same technology frameworks which created innovative services and business solutions have shortened the time taken to change these innovative value-added services to commodity, thus, they lose their value-add to the customers.

This phenomena is a reflection of the much shorter cycle of technology, service and business innovations. This accellerates the competition, which then further feeds to this cycle. This poses two serious challenges to the telco industry.

First, while the technology, service, and business innovations continue to drive and intensify the competition, the customer management becomes the defining point for the meaning of competition. With more and more telecommunication, services and products become commodity, and innovation in achieving customer intimacy and managing customer relationships becomes crucial. This innovation will give a service provider more advantage in competition than the value-add in the services themselves.

Second, while this quick and short innovation cycle has put so much demand on the underlying technology framework for support, the telcos cannot, however, afford the major business process reengineering any more, particularly the major reengineering exercises with a discontinuous and disruptive nature. This means that while the technological evolution cycle is getting shorter and more dynamic, the business solution life cycle is getting longer and more stable. This reflects telco's desire to benefit more and spend less on the BPR.

To successfully confront these challenges requires a stronger vision and sound strategy in software development efforts to support the continuous improvement of customer services and business innovation, yet to require less frequent business process reengineering. The business solution life-cycle management principles, processes, and experiences discussed in this book should become more important in the new era of telecommunication industry.

Glossary

ACID Atomicity, Consistency, Isolation and Durability (Transaction properties)
ACSE Association Control Service Element
ADSL Asymmetric Digital Subscriber Line
API Application Programming Interface
ASN.1 Abstraction Syntax Notation 1
ATM Asynchronous Transfer Mode
ATMF ATM Forum
BPR Business Process Reengineering
CCITT Consultative Committee of International Telegraph and Telephone (now ITU-T)
CER Call Event Record
CLEC Competitive Local Exchange Carrier
CMIP Common Management Information Protocol
CMIS Common Management Information Service
CORBA Common Object Request Broker Architecture
COSS Common Object Service Specification
CTI Computer Telephony Interface
DCE Distributed Computing Environment
DCF Data Communication Function
DCN Data Communication Network
DCOM Distributed Component Object Model
DN Distinguished Name
DPE Distributed Processing Environment (defined by TINA-C)
DSF Directory System Function
DTP Distributed Transaction Processing
EFD Event Forwarding Discriminator
EML Element Management Layer
EMS Element Management System
ETSI European Telecommunications Standards Institute

FAB Fulfillment, Assurance and Billing
FCAPS Five TMN management functions (Fault, Configuration, Accounting, Performance and Security)
FCC Federal Communication Commission
FSA Finite State Automata
GDMO Guidelines for the Definition of Management Objects (OSI)
GNM Generic Network Model
GOM Generic Object Model
GRM General Relationship Model
GUI Graphical User Interface
HTML Hypertext Markup Language
ICF Information Conversion Function
IDL Interface Definition Language
I-ETS Interim European Telecommunication Standards
IIOP Internet Inter-ORB Protocol
ILEC Incumbent Local Exchange Carrier
ISO International Standards Organization
ISV Independent Software Vendor
IT Information Technology
ITU-T International Telecommunications Union—Telecommunications Standardization Sector (formerly CCITT)
IVR Interactive Voice Response
JIDM Joint Inter Domain Management (X/Open and TMF)
MAF Management Application Function
MD Mediation Device
MF Mediation Function
MIB Management Information Base
MO Managed Object
NE Network Element
NEF Network Element Function
NM Network Management
NMF Network Management Forum (now TeleManagement Forum)
NML Network Management Layer
NMO Network Management Objects
NMS Network Management System
NNM Network Node Manager
OA&M Operation, Administration and Maintenance
ODBC Open Database Connectivity
ODBMS Object Database Management System
ODL Object Definition Language (defined by TINA-C)
ODMA Open Distributed Management Architecture
OLTP On-Line Transaction Processing
OMG Object Management Group
ORB Object Request Broker
OS Operation Systems
OSF Open Software Foundation
OSI Open System Interconnection

OSS	Operation Support System
OTS	Object Transaction Service
Q3	TMN communication protocol
QA	Q Adapter
QAF	Q Adapter Function
QoS	Quality of Service
RDBMS	Relational Database Management Systems
RDN	Relative Distinguished Name
RFC	Request for Comments
RFI	Request for Information
RFP	Request for Proposal
ROSE	Remote Operation Service Element
RPC	Remote Procedure Call
SDH	Synchronous Digital Hierarchy
SLA	Service Level Agreement
SM	Service Management
SMC	System Management Communication
SMF	System Management Functions
SMI	System Management Information
SMO	Service Management Object
SNMP	Simple Network Management Protocol
SONET	Synchronous Optical Network
SQL	Structured Query Language
STM-N	Synchronous Transfer Module level N
TCP/IP	Transmission Control Protocol/Internet Protocol
TIM	Technology Integration Map (TM Forum)
TINA-C	Telecommunication Information Network Architecture—Consortium
TL1	Translation Language-1
TMF	TeleManagement Forum (formerly Network Management Forum)
TMN	Telecommunications Management Network
TOM	Telecom Operations Map (TM Forum)
TP	Transaction Processing
UML	Unified Modeling Language
VCN	Virtual Circuit Number
VOD	Video On Demand
VP	Virtual Path
VPN	Virtual Private Network
WDM	Wavelength Division Multiplex
WS	Workstation
WSF	Workstation Function
XMP	X/Open standards for Management Protocols API
XOM	X/Open standards for OSI Object Management API

References

[1] E. K. Adams and K. J. Willetts, When service management is your business, *Journal of Network and Systems Management,* Vol. 4. No. 4. pp. 349–353, 1996.

[2] E. K. Adams and K. J. Willetts, *The lean communications provider—Surviving the shakeout through service management excellence,* New York: McGraw-Hill, 1996.

[3] S. Aidarous and T. Plevyak (eds.), Telecommunications network management into the 21st century: Techniques, standards, technologies, and applications, Piscataway, NJ: IEEE Press, ISBN 0-7803-1013-6, 1994.

[4] S. Aidarous and T. Plevyak (eds.), Telecommunications network management technologies and implementations, Piscataway, NJ, IEEE Press, ISBN 0-7803-3454-X, 1998.

[5] M. Bauer, Advances in systems and application management: Managing large distributed computing environments, *Journal of Network and Systems Management,* Vol. 4, No. 4, pp. 477–480, December 1996.

[6] A. Bedoyan and A. Clemm, Experiences with the integration of a legacy management system into a TMN environment, *Proceedings of IEEE/IFIP 1998 Network Operations and Management Symposium,* New Orleans, LA, USA, February 1998.

[7] L. Bernstein and C. M. Yuhas, Plug and play TMN: Rocky road to a bright hope, *Journal of Network and Systems Management,* Vol. 3, No. 1, pp. 9–12, March 1995.

[8] L. Bernstein, E. Bryan, and C. M. Yuhas, The system is the business—But at what cost? *Journal of Network and Systems Management,* Vol. 4, No. 1, March 1996.

[9] C. J. Byrne, L. G. Raman, and H. Woo, Realizing a TMN, *Journal of Network and Systems Management,* Vol. 3, No. 1, March 1995, pp. 39–72.

[10] R. Chadha and S-Y. Wu, Incorporating manageability into distributed software, in *Integrated management in a virtual world,* A. Lazar, P. Saracco, and R. Stadler (eds.), London: Chapman & Hall, 1997.

[11] T. Chatt, J. Sepp, M. A. Curry, and U. Hollberg, TMN/C++: An object-oriented API for GDMO, CMIS, and ASN.1, in *Integrated management in a virtual world,* A. Lazar, P. Saracco, and R. Stadler (eds.), London: Chapman & Hall, 1997.

[12] G. Chen, Distributed transaction processing standards and their applications, *Computer Standards and Interfaces* 17 (1995) pp. 363–373.

[13] G. Chen, M. Neville, and Q. Kong, Distributed network management using CORBA/TMN, *Proceedings of 7th IFIP/IEEE International Workshop on Distributed Systems: Operations and Management,* L'Aquila, Italy, October 1996.

[14] G. Chen and Q. Kong, Integrated TMN service provisioning and management environment, in *Integrated management in a virtual world,* A. Lazar, P. Saracco, and R. Stadler (eds.), London: Chapman & Hall, 1997.

[15] G. Chen, Building integrated TMN solutions using multiple technologies, *NMF Bulletin,* Spring issue, 1997.

[16] G. Chen and Q. Kong, The business process and object modelling for service ordering, *Proceedings of the 8th International Workshop on Distributed Systems and Operations and Management* (DSOM), A. Seneviratne, V. Varadarajan, and P. Ray (eds.), October 1997.

[17] G. Chen, J. Rixon, and Q. Kong, Integrating CORBA and Java for ATM connection management, *Proceedings of the 8th International Workshop on Distributed Systems: Operations and Management* (DSOM), A. Seneviratne, V. Varadarajan, and P. Ray (eds.), October 1997.

[18] G. Chen, *Building integrated TMN solutions: From OSI to CORBA,* Tutorial given on 1998 Network Operations and Management Symposium, New Orleans, LA, USA, February 1998.

[19] G. Chen, Q. Kong, and P. Benain, SP-SP service ordering specification and its implementation, *Proceedings of the 6th Network Operations and Management Symposium,* New Orleans, LA, USA, February 1998. ISBN 0-7803-4351-4.

[20] G. Chen, Generic object model, *CiTR technical white paper,* 1998.

[21] G. Chen, War and peace—How to avoid technology wars? *Journal of Network and Systems Management,* Vol. 6, No. 3, September 1998.

[22] G. Chen, Service differentiation through electronic customer access in a competitive telco environment, Proceedings of IEICE Workshop, Japan, 1998.

[23] G. Chen, Q. Kong, J. Etheridge, and P. Foster, Integrated TMN service management, to appear in *Journal of Network and Systems Management,* 1999.

[24] J. P. Chester and K. R. Dickson, Standards for integrated services and networks, in *Integrated Network Management IV,* A. S. Sethi, Y. Raynaud, and F. Faure-Vincent (eds.), London: Chapman & Hall, 1995.

[25] B. Couch, Process flow-through for service provisioning, *Proceedings of 1998 Network Operations and Management Symposium,* New Orleans, LA, USA, February 1998.

[26] A. Dittrich and M. Hoft, Integration of a TMN-based management platform into a CORBA-based environment, *Proceedings of 1996 Network Operations and Management Symposium,* Kyoto, Japan, April 1996.

[27] J. Etheridge, G. Chen, M. Tanaka, and G. Watanabe, Distributed architecture for cross-domain network management, *Proceedings of the 6th Network Operations and Management Symposium,* New Orleans, LA, USA, February 1998. ISBN 0-7803-4351-4.

[28] M. Flauw and P. Jardin, Designing a distributed management framework—An implementer's perspective, *Integrated Network Management IV,* A. S. Sethi, Y. Raynaud, and F. Faure-Vincent (eds.), London: Chapman & Hall, 1995.

[29] N. Furley, OSS architecture framework, *Proceedings of 1996 Network Operations and Management Symposium,* Kyoto, Japan, April 1996.

[30] G. Genilloud and L. Fang, Integrating the OSI and the distributed objects architectures for system management, *Proceedings of 1998 Network Operations and Management Symposium,* New Orleans, LA, USA, February 1998.

[31] D. Georgakopoulos and M. Hornick, An overview of workflow management: From process modelling to workflow automation infrastructure, *Distributed and Parallel Databases,* Vol. 3, pp. 119–153 (1995).

[32] D. Griffin and G. Pavlou, Realizing TMN-like management services in TINA, *Journal of Network and Systems Management,* Vol. 5, No. 4, December 1997.

[33] T. Hamada, H. Kamata, and S. Hogg, An overview of the TINA management architecture, *Journal of Network and Systems Management,* Vol. 5, No. 4, December 1997.

[34] M. Hammer and J. Champy, *Re-engineering the corporation: a manifesto for business revolution,* NSW, Australia: Allen & Unwin, 1994.

[35] J. W-K. Hong, J-S. Kim, and J-K. Park, A CORBA-based quality-of-service management framework for distributed multimedia services and applications, *IEEE Network,* Vol. 13, No. 2, pp. 70–79, 1999.

[36] J. W-K. Hong, A CORBA-based distributed multimedia collaboration system, *Advances in Distributed Multimedia Systems,* Edited by S. Vuong, Singapore: World Scientific Publisher, Spring 1999.

[37] M. Hsu and M. Howard, Workflow and legacy systems, *BYTE,* July, 1994.

[38] H. Inamori, Common software platform for realizing a strategy for introducing the TMN, *Proceedings of 1998 Network Operations and Management Symposium,* New Orleans, LA, USA, February 1998.

[39] K. Iseda, T. Chuja, and T. Suzuki, CORBA-based network operation system architecture, *Proceedings of 1998 Network Operations and Management Symposium,* New Orleans, LA, USA, February 1998.

[40] H. Ishikawa, T. Oimatsu, and N. Akiyama, Data communication platform in distributed operations system based on TMN, *Proceedings of 1996 Network Operations and Management Symposium,* Kyoto, Japan, April 1996.

[41] P. A. Johnson, The refocusing of TMN and beyond, *Journal of Network and Systems Management,* Vol. 3, No. 1, pp. 139–143, March 1995.

[42] Q. Kong, G. Chen, and G. Holliman, Telecommunication service management, *Data Communications OpenView Advisor,* Vol. 1, No. 8, August 1995.

[43] Q. Kong and G. Chen, Integrating CORBA and TMN environments, *Proceedings of the 1996 Network Operations and Management Symposium,* Kyoto, Japan, April 1996.

[44] Q. Kong and G. Chen, Transactional workflow in telecommunication service management, *Proceedings of the 1996 Network Operations and Management Symposium,* Kyoto, Japan, April 1996.

[45] Q. Kong, G. Chen, and N. Lawler, Distributed architecture for IN service management, *Proceedings of the IEEE Intelligent Network Workshop* IN'96, Melbourne, April 1996.

[46] Q. Kong, G. Chen and R. Hussain, A management framework for internet services, *Proceedings of the 6th Network Operations and Management Symposium,* New Orleans, LA, USA, February 1998. ISBN 0-7803-4351-4.

[47] G. E. Krasner and S. T. Pope, A cookbook for using the Model-View-Controller user interface paradigm in Smalltalk-80, *Journal of Object-Oriented Programming,* SIGS Publications, New York, Vol. 1, No. 3, pp. 26–49, 1988.

[48] W-C. Lee, A framework for TMN-CORBA interoperability, *Proceedings of 1998 Network Operations and Management Symposium,* New Orleans, LA, USA, February 1998.

[49] D. Lewis, T. Tiropanis, R. Shi, and A. Richter, A service engineering approach to inter-domain TMN systems development, in *Integrated management in a virtual world,* A. Lazar, P. Saracco, and R. Stadler (eds.), London: Chapman & Hall, 1997.

[50] J. Lin, S. Liu, M. Azuma, and M. Ejiri, TMN based network management systems in a distributed environment, *Proceedings of 1998 Network Operations and Management Symposium,* New Orleans, LA, USA, February 1998.

[51] H. Mulder and J. Pavon, TINA: Evolving the TMN, *Journal of Network and Systems Management,* Vol. 5, No. 4, December 1997.

[52] B. Muller, A discussion of ILEC/CLEC interconnection to OSS systems, *Proceedings of 1998 Network Operations and Management Symposium,* New Orleans, LA, USA, February 1998.

[53] J. L. Oliveira and J. A. Martins, A management architecture based on network topology information, *Journal of Network and Systems Management,* Vol. 2, No. 4, December 1994.

[54] G. Pavlou, K. McCarthy, S. Bhatti, and G. Knight, The OSIMIS platform: Making OSI management simple, *Integrated Network Management IV,* Adarshpal S. Sethi, Yves Raynaud, and Fabienne Faure-Vincent (eds.), London: Chapman & Hall, 1995.

[55] C. Peace, Business process for automated customer service, *Proceedings of 1996 Network Operations and Management Symposium,* Kyoto, Japan, April 1996.

[56] A. Pope, The CORBA reference guide: Understanding the common object request broker architecture, Reading, MA: Addison-Wesley, ISBN 0-2016-3386-8.

[57] S. Rahkila and S. Stenberg, Experiences on building a distributed computing platform prototype for telecom network and service management, in *Integrated management in a virtual world,* A. Lazar, P. Saracco, and R. Stadler (eds.), London: Chapman & Hall, 1997.

[58] I. Rose, CORBA-at-large, *CiTR technical white paper,* CiTr, Queensland, Australia: 1997.

[59] J. Rumbaugh, M. Blaha, W. Premerlani, F. Eddy, and W. Lorensen, *Object-oriented modeling and design,* Englewood Cliffs, NJ: Prentice Hall, 1991.

[60] J. K. Shrewsbury, An introduction to TMN, *Journal of Network and Systems Management,* Vol. 3, No. 1, pp. 13–38, March 1995.

[61] J. Siegel, CORBA fundamentals and programming, John Wiley & Sons, ISBN 0-4711-2148-7, 1996.

[62] M. Sloman, Policy driven management for distributed systems, *Journal of Network and Systems Management,* Vol. 2, No. 4, December 1994.

[63] N. Soukouti and U. Hollberg, Joint inter domain management: CORBA, CMIP and SNMP, in *Integrated management in a virtual world,* A. Lazar, P. Saracco, and R. Stadler (eds.), London: Chapman & Hall, 1997.

[64] G. Spinelli, G. Martini, W. Takita, and S. Chikara, TINA connection management implementation over distributed processing platforms, *Proceedings of 1996 Network Operations and Management Symposium,* Kyoto, Japan, April 1996.

[65] L. Strick, M. Wittig, and A. Berquist, A framework for designing distributed management applications, *Journal of Network and Systems Management,* Vol. 6, No. 2, June 1998.

[66] M. Tanaka, K. Arima, H. Hayashi, and S. Fukui, Platon: The distributed NMS platform—with CORBA/TMN integration, *NEC Research & Development,* Vol. 38, No. 3, July 1997.

[67] J. S. Tjong, Implementation of a CMIP-based network management system, *Journal of Network and Systems Management,* Vol. 3, No. 1, pp. 73–83, March 1995.

[68] M. Tschichholz, J. Hall, S. Abeck, and R. Wies, Information aspects and future directions in an integrated telecommunications and enterprise management environment, *Journal of Network and Systems Management,* Vol. 3, No. 1, pp. 111–138, March 1995.

[69] T. Uslaender, User requirement and state-of-the-art in managing CORBA-based applications, *Journal of Network and Systems Management,* Vol. 5, No. 4, December 1997.

[70] R. Whitner, Designing scalable applications using CORBA, in *Integrated management in a virtual world,* A. Lazar, P. Saracco, and R. Stadler (eds.), London: Chapman & Hall, 1997.

[71] E. D. Zeisler, TMN CORBA matures: a service provider gateway for measuring service availability, *Proceedings of 1998 Network Operations and Management Symposium,* New Orleans, LA, USA, February 1998.

Standard Specifications

[72] ATM Forum Technical Committee, *M4 Network-View Interface Requirements and Logical MIB,* AF-NM-0058.000, March, 1996.

[73] ETSI, I-ETS 300 653: Telecommunication Management Network (TMN)—Generic Managed Object Class Library for the Network Level View, March 1996.

[74] ISO 9595, ITU-T X.710, *Information Technology—Open Systems Interconnection—Common Management Information Service Definition (CMIS),* October, 1997.

[75] ISO 9596-1, ITU-T X.711, *Information Technology—Open Systems Interconnection—Common Management Information Protocol Specification (CMIP),* October, 1997.

[76] ISO 10026-1, *Information Technology—Open Systems Interconnection Distributed Transaction Processing, Part 1 OSI TP Model,* 1992.

[77] ISO 10164-1, (ITU-T X.730), *Information Technology—Open Systems Interconnection—Systems Management: Object Management Function,* 1992.

[78] ISO 10164-5, (ITU-T X.734), *Information Technology—Open Systems Interconnection—Systems Management: Event report management function,* 1992.

[79] ISO 10165-4, ITU-T X.722, *Information Technology—Open Systems Interconnection—Structure of Management Information, Part 4: Guidelines for the Definition of Managed Objects (GDMO),* January 1992.

[80] ISO 11587, ITU-T X.702, *Information technology—Open Systems Interconnection—Application context for systems management with transaction processing,* November 1995.

[81] ISO 13244, ITU-T X.703, *Information technology—Open Distributed Management Architecture,* October 1997.

[82] ITU-T Recommendation M.3010: *Principles for a telecommunication management network,* May 1996.

[83] ITU-T M.3100: *Generic network information model,* July, 1995.

[84] ITU-T M.3200: *TMN Management Services and telecommunications managed areas: overview,* April 1997.

[85] ITU-T M.3400: *TMN management functions,* April 1997.

[86] ITU-T G.805, *Generic Functional Architecture of Transport Networks,* November 1995.

[87] ITU-T X.903, *Information technology—Open distributed processing—Reference model: Architecture,* November, 1995.

[88] OMG, *The Common Object Request Broker: Architecture and Specification,* Revision 2.0, December 1995.

[89] OMG, *CORBA services: common object services specification,* December 1998.

[90] OMG, *JIDM Interaction translation,* Final Submission to OMG's CORBA/TMN Interworking RFP, OMG Document: Telecom/98-10-10, 1998

[91] OMG, Telecom Task Force, *CORBA-based telecommunication network management system,* OMG White Paper, May 1996.

[92] TINA-C, *Service architecture,* Version 5.0, Public release, June 1997.

[93] TINA-C, *Requirements upon TINA-C architecture,* Version 2.0, February 1995.

[94] TINA-C, *TINA business model and reference points,* Version 4.0, May 1997.

[95] TINA-C, *TINA network resource architecture,* Version 3.0, February 1997.

[96] TINA-C, *TINA distributed processing environment (TINA-DPE),* December 1995.

[97] TM Forum, *A service management business process model,* 1995.

[98] TM Forum, *TMN/C++, application programming interface,* NMF, Issue 1.0, 1996

[99] TM Forum, *SmartTMN Telecom operations map,* Evaluation Version Release 1.0, TMF GB910, October 1998.

[100] TM Forum, *SmartTMN Technology integration map,* Issue 1, TMF GB909, April 1998.

[101] X/Open, Guide G307: *Distributed transaction processing reference model* (V2), X/Open Company Ltd., November 1993.

[102] X/Open, *Inter-domain management: specification translation,* X/Open Preliminary Specification P509, ISBN 1-85912-150-0, March 1997.

Index

F

G

H

I

P

R

S

U

W

About the Authors

Graham Chen is currently the chief technologist of CiTR. He received the Ph.D. in computer science from Heriot-Watt University, Edinburgh, U.K., in April 1986 and the B.Sc. degree in computer science from Nanjing University, China, in January 1982.

Dr. Chen has an extensive software development, R&D management and business/technology consulting experience and, in the past 10 years, in the telecommunication network and service management and distributed systems area. He is responsible for CiTR's technology and product strategy and leads CiTR's Advanced Technology Development program; he has worked on a large range of commercial product developments, software solution development projects, and consulting projects for CiTR's major clients.

An active member of the international research community, Dr. Chen is a regular speaker and committee member in a number of leading international conferences in the network management area. Also, he is a member of IEEE Computer Society and IEEE Communications Society.

Qinzheng Kong is currently a senior consultant and software architect of CiTR. Dr. Kong received the Ph.D. in computer science from Heriot-Watt University, Edinburgh, U.K., in December 1989. She received the M.Sc. degree in 1985 and the B.Sc. degree in 1982, both of which are in computer science from Nanjing University, China.

Dr. Kong has a wide range of software development and business/technology consulting experience and, in the past 10 years, in the Telco network and service management and distributed systems area. She is a major contributor to CiTR's technology and product strategy and CiTR's Advanced Technology Development program, and has worked in a large range of commercial product development, solution projects, and consulting projects for CiTR's major clients. She has solid commercial-software development experience in TMN solutions, OSL, and distributed systems.

Dr. Kong has published widely in leading international conferences and journals. Her work in CORBA/TMN integration was well recognized as one of the earliest contributions in the industry.